The
CNC
ToolBox

Top Service for Machine Tools
2nd Edition, Revised

The CNC ToolBox

Top Service for Machine Tools
2nd Edition, Revised

Daniel D. Nelson

Aero Publishing  Bristol, WI

Published by:　　Aero Publishing
PO Box 579
Bristol, WI 53104 U.S.A.

Copy Edit by Nancy Cherrington and Jonathan Woods.
Cover Design and Artwork by Larry A. Nelson.

© 1996 and 1999 by Daniel D. Nelson
First Printing 1996
Second Printing 1999, revised

Publisher's Cataloging-in-Publication
(Provided by Quality Books, Inc.)

Nelson, Dan (Daniel D.)
　　The CNC toolbox : top service for machine tools
/ Daniel D. Nelson. — 2nd ed., rev.
　　p. cm.
　　Includes bibliographical reference and index.
　　LCCN: 99-72277
　　ISBN: 0-9654314-7-9

　　1. Machine-tools—Numerical control.　I.
Title.

TJ1189.N43 1999　　　　　621.9'023
　　　　　　　　　　　　QBI99-500147

Warning and Disclaimer

The author and publisher have taken care in preparation of this book, but there may be omissions and inaccuracies as to specific types, groups, and manufacturers of machine tools, and other related services and equipment. The author and/or Aero Publishing cannot and do not warrant the accuracy, completeness, correctness, currentness, or fitness for a particular purpose of the information contained in this book. The author and/or Aero Publishing shall have neither liability nor responsibility to any person or entity with respect to any loss or damage, or alleged to be caused, directly or indirectly by the information contained in this book.

It is not the purpose of this book to reprint all the information available to the author and/or publisher, but only to educate and entertain. It remains your responsibility to exercise care and normal precautions to prevent personal injury and equipment damage. If expert assistance is required, the services of a competent professional should be sought.

Overview

Contents

1st Tool–*Knowledge*

Servo Loop

Spindle Loop

Control Features

Mechanical Features

Optional Features

2nd Tool–*Diagnostics*

Test Equipment

Electric Power

Servo Loop

The External Signals

The Internal Signals

Spindle Loop

Computer Numerical Control

Overall System Features

3rd Tool–*Problem Solving*

Three Skills for Service

Game Plan

Solving Problems

Introduction

A revolution in machine tool technology occurred during the late seventies. Affordable, high quality CNC (Computer Numerical Control) machines were introduced into the open industrial marketplaces of the world. The fortunate few to purchase CNCs back in the '70s reaped a steady return of manufacturing profit. In fact, many of the established shops of today owe their existence to a few dependable early CNC machines.

People in the industry first heard about new computer controlled machines in the late '70s, began buying them in numbers in the '80s, and to this day are looking for ways to increase their productivity using them. At the start, people became concerned about computers and robots taking over people's jobs in machine shops. The five guys standing in front of five manual engine lathes were quickly replaced by one guy pushing a green button on the new computer controlled machines. However, now that 25 years have passed the demand for challenging, highly-skilled technical jobs has returned from this machine tool revolution.

The technology used to run computer controlled machines is complex and changes fast. Understanding common CNC control concepts is surprisingly easy, the difficulty comes in fitting basic control concepts with the sparse collection of cryptic documents included with each machine. This is made considerably more challenging by the fact that machines are built worldwide by numerous makers in countless configurations.

At one time each and every NC machine was brand new, sitting in a dealer's show room, with eager sales people announcing this latest in motion control technology. When financing was arranged, the machine was sold and delivered by semi-tractor trailer to an expectant end-user's shop. After being fork lifted into place and quickly set up, the year-after-year working lifetime of the machine began. So where are all these NC machines today?

Of course, the low quality tools are in the junk heap, but those amazing "quality" machine tools, built anytime in the last twenty five years, are still running strong. These machines are always sold off to someone else willing to put them back in service. This is the secret of quality possessed by some of these mighty machines: faithfully turning out parts inside residential garages, sharing space with the family station wagon, or running at large corporations, placed in side-by-side, expanding cellular formations.

No two machine shops in the field are exactly the same, whether a machine is running three shifts, roughing out iron castings in a smoke-filled hot house, or in a linoleum floored hospital supply company, the same requirements apply: a machine in production makes good money; when the machining motion goes down, the money abruptly stops.

When a machine goes down, the news travels fast. Perhaps an entire car assembly line waits for aluminum cylinder blocks. The cylinder blocks are in ceiling-high piles waiting for that single machining operation normally performed by the down NC machine. Unfortunately this is the machine for the job—it contains the critical jigs, tooling and tested part program. The repercussions begin to travel.

The part contracts must be met. Maybe the job is moved to another machine, farmed out to another company, or lost altogether. All the top guns are counting on one thing happening: getting it fixed! As soon as possible, with a minimum of mistakes. This is the never ending nature of NC service.

Preface

The computer controlled machinery—known in machining parlance as CNCs—offer an excellent potential for profit. Attracted by the profit is a large supporting cast of eager business professionals. All of them gainfully employed using their skill and insight to the technology of these *computer numerically controlled* machine tools.

The average reader interested in computer controlled machines will find practical explanations for how the machines work, where to find outside support and finally what it takes to get a machine properly serviced. The same standards and procedures that factories use to build these machines apply to making repairs in the field. Valuable information for anyone owning, operating, servicing or selling these machines.

The goal of this book is to present CNC technology in a format that captures the subject of **all** CNC machines, not just a single proprietary system. To this end, the reader is presented *clear air*, a common body of information for CNC machine tools slowly revealed in the course of visiting hundreds of the well-run, well-maintained shops in the field. The *hazy clouds* of divergent, service-numbing details are avoided, this highly specific information must still come from the individual factories and service outfits. They have the access, experience and most important the **responsibility** for supporting their relatively narrow applications of single-brand, single-type machine tools. Their responsibility is to fix it; understanding how they fulfill their responsibility is what this book is all about.

I heard some wild stories and met many good people while performing machine tool service across North America. The endless goings-on from the service years, taking phone calls and sorting out countless on-site visits, demonstrated the need and gave inspiration for the first and second editions of this book.

Daniel D. Nelson

1st Tool

Knowledge

Introduction

Part One of this book focuses on understanding the basic business of servicing a CNC tool. Coverage is concentrated on those ideas that originate and recur most often in the field. These ideas build a base for the testing Diagnostics and the Problem Solving tools that follow in Parts Two and Three.

The discussion begins with the established maintenance network for machine tools. This service network includes all the players participating in the business, the big CNC-makers, importers, dealers, independent service companies and final end-users.

The issue of personal safety is emphasized early on in this first section and throughout the text. The service business demands an enlightenment to the endless lists of safety hazards. This subject concerns everyone who has heard the grizzly stories—and should especially admonish the novice.

To conclude this section, direct knowledge is presented for the sub-systems and components incorporated into many CNC machines. These discussions focus on the machine itself, how discreet components like motors, inverters, computer systems, mechanical systems and so on, all act in unison to provide the basic function of CNC-tools: converting electrical power into highly controlled mechanical power.

Keep in mind that additional details are always available. Reference to factory publications, OEM phone support, training systems, outside reading and on-site service visits are recommended throughout the text.

1 *Maintenance Network*

1.1 Introduction

Machine tool consumption in the United States annually exceeds eight billion dollars,[1] with some 60 billion dollars in sales world-wide. Each quality machine sold receives, on average, twenty years of field maintenance support. Because the number of machines in the field keeps climbing, service of CNC machines is a growing business.

The new machines are manufactured with the latest technology to reduce cost while expanding reliability and performance. The older machines occasionally receive factory re-engineering to make available a fresh supply of modern replacement parts.

The world-wide model for marketing and service support of machine tools is best described by a network of services. To understand this *maintenance network*, follow the money trail. Cash starts with the *end-user* and ends with the big CNC-tool factories—known as a group as OEMs, or original equipment manufacturers. Along this winding cash path, the machine importers, machine dealers and assorted service companies are all positioned.

Each group has responsibility for different types of maintenance support. After a new machine purchase, all the pieces in the maintenance network should fit into total support for the end-user. Finding the best route to follow through the network for older, out of warranty machines is currently growing in new directions, some away from traditional boundaries.

1.2 Machine Dealers

Buying a machine from an established machine dealer provides immediate membership into the maintenance net-

---

1 The Association for Manufacturing Technology and the American Machine Tool Distributors Association report that US manufacturers' consumption of industrial machinery and equipment in 1997 totaled $8.6 billion.

work. A machine tool dealer will bend over backwards to find the answer to a customer's question. If a question comes up they cannot answer, they call on other diligent people around the network who have good reputations for quickly supplying the key information. A satisfied customer will in turn provide the dealer with repeat business and sales commissions.

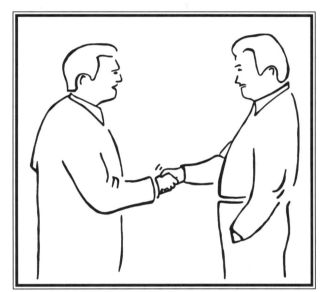

Hand Shake

Primarily, a dealer's responsibility is to keep machines in production, questions concerning machine application and programming are generally their strongest suit. Experienced with the machines they are selling, they can get prompt action from the control or machine builder if something goes wrong.

Exceptional dealers offer total support to the customer. Better established dealers even offer *turnkeys*, a top-to-bottom approach to selling a machine in which everything is included in one low, low price. Everything is included; from the tooling, cycle-time simulations and programming, right down to on-site scheduling and management of visiting service engineers.

Simple questions for a customer to ask a dealer can include: Have you a field service department? Do you offer turnkeys? Are there many people in the office? Do they specialize in this type of machine? Have you sold this line of machinery a long time? Will you come over if I have any problems? Plenty of machine tool dealers can honestly answer "yes" to all of those questions. Established machine shops maintain close working relationships with all their dealers.

The dealer's approach to service on older machines is a careful one. Long-standing, mysterious or expensive no-win problems are not inviting unless the costs can somehow be defrayed. Everyone in service aims to be helpful, but the machine tool business is like any other business—occasionally unpopular decisions stand firm.

Any problem is solvable—for a price. An independent service company, a dealer, the control builder and the machine builder may all make a service visit before a final, elusive answer is found. The customer is expected to pay for all the service labor and materials. However, final service charges are usually held up until the machine is actually running and then some friendly negotiations may occur. In the end there is often more than just one customer. Strictly speaking the OEMs show up for the dealers and the dealers actually cater to the end-user, but a final negotiation for the services rendered can occur on many levels.

1.3   Control Builders

The major league control builders maintain a world-wide presence of satellite offices staffed with factory trained field service engineers. These factory sponsored facilities provide assistance to machine builders, dealers and end-users anywhere in the world. If *control-side* problems are suspected, an effort to locate the problem is usually offered free of charge over the telephone.

Replacement parts, diagnostic checking and specialized documents are provided to the customer. When problems cannot be solved over the telephone, a service call on-site is arranged. The strictly machine-side and application problems are generally referred to the responsible machine builder or dealer for investigation. Factory referrals are dependent upon what combination of builders are involved.

Electronic replacement parts are sold on a straight *sales* or returnable *exchange* basis to almost anyone who calls the service department.

A straight part sale means the part goes out and doesn't come back. Examples include fuses, fans, motors and power transistors, any items which cannot be returned safely to the warehouse shelf. Some discretion with this policy is applied to parts returning from a managed service investigation. In recent years, replacement electrical parts have become dramatically more expensive. When replacement control parts are needed, a good preliminary diagnosis avoids costly mistakes.

The exchange part program starts with good parts sent to a customer. If they fix the machine, the original defective parts are boxed up and returned for an exchange, or core credit. Return parts are tested, repaired, and with factory approval, passed on to the next customer. The credit posted upon return of an unused exchange part is based on each builder's specific service policy. A fixed percentage "restocking" charge for unused parts is common in the business.

Parts availability is becoming an important issue for the older machines. The new factory production of most CNC-parts ends after only a few years. What remaining stock of parts that are left will eventually be used by sales and exchange parts programs. When parts become out of stock, the remaining factory option is to send defectives directly to repair centers for a *customer property* repair. Service vendors know what is actually available and what is repairable, the decision to keep, retrofit or sell a machine may depend on the results of having a frank conversation with the vendor on this issue.

The control builders' business is selling new motion control electronics. Critical to this mission is a good reputation for field service.

1.4   Published Documents

During a machine tool's initial design and construction, the factories compile a host of related service documents. A basic set of these books, along with a small box of tools and perhaps some replacement fuses, are included with each new machine.

The primary training objective of the factory manuals is programming the machine to make parts. The available maintenance books, schematics and related diagrams are limited and difficult to understand. It seems that every topic raised introduces an ever-growing area of additional studies without specific reference or follow-up assistance.

The typical service documents included from the factory are summarized in Appendix One. Contact the OEM directly for a comprehensive list of documents published for a given machine tool. If certain documents are missing, copies are obtainable from the dealer or directly from the factory. Every machine needs a complete set.

The original factory documents are valuable, with some of the older ones getting especially hard to find. Sometimes a photocopied set from the OEM is the best option for older machines. Successful service often comes down to keeping a complete, readable set of factory books with the machine.

1.5   Specialized Documents

The longer a machine tool is running in the field, the better developed the field service information becomes. This is especially true if thousands of the same control computer and machine type are sold and serviced.

A seasoned factory engineer receives the same questions every day. Common service problems have already been heard—likely a hundred times. To make such repetitive service jobs easier, the *specialized* service documents are written. Verbally repeating the same repair procedures over the phone, every day, becomes very tedious. Also, communication errors can creep in and screw up an otherwise routine repair.

Specialized maintenance documents are routinely faxed after a brief consultation with an OEM service engineer. Some procedures are handwritten, while others are stored and created on desktop *PC*s (Personal Computers). Many versions of the same thing can exist, depending on who is contacted. Any of this valuable, specialized information for a machine should be kept with the machine books after serving the immediate service purpose.

Some OEMs offer convenient *fax-back* numbers for the frequently requested documents. Appendix Two lists some of the common topics detailed by these specialized maintenance documents. A well-prepared document speeds up a machine service and reduces costly service surprises.

1.6   Future Support Systems

Machine tool information is becoming available on the information superhighway.[2] Gaining access to existing on-line systems requires having a personal computer, a high speed fax-modem and an account with an on-line service or Internet access company. Once the PC-modem is installed, access to specialized maintenance documents, factory publications, computer based testing and other "knowledge tool" type information over the telephone line is a snap. Asking an OEM

---

2 Surf the web to popular CNC sites like SME.org, mmsonline.com, mfg-net.com, or Penton.com/am. Find books at Amazon. com Find a list of good CNC books and the key references for this book "on-line" at http://www.cncbookshelf.com

about their latest improvements in on-line systems increases the perceived value of offering such services.

Another interesting development is the *CBT* (Computer Based Testing) and "expert" systems for CNC service. These systems store and distribute knowledge about highly specific products and applications using scanned factory documents placed in logical order using an index. The moniker "expert system" applies to the new software strategies for adding interactive diagnostics and creative problem solving, in conjunction with stored knowledge.

As the competition for service heats up, savvy customers are finding more options for their business. Advertisements in the back of the trade magazines peddle electronic and mechanical parts for CNC machine tools that were previously only available from the OEMs.

1.7   Machine Tool Builders

The machine's backbone iron casting is developed at the machine builder's factory. Fitted to this main casting is the tool changer, turrets, pumps, ball screws, spindle bearings and sheet metal covers.

After a machine builder decides on the basic size, weight and configuration of a new machine, an analysis of application and control formulae supply the solutions for speed and acceleration of a new machine tool. A visit from the control builder decides the appropriately-sized motors and drives for the new machine. Larger motors and servo drives are naturally selected as the machine grows in size. The same internal NC control computer is easily reprogrammed to handle different size machines.

Questions concerning the machine's mechanicals are best answered by the original machine tool builder. Mechanical replacement parts, mechanical drawings and adjustment specifications are readily available. On-site service calls are arranged for the tougher problems and service evaluations. A visiting serviceman from the machine builder carries the heavy tool box and offers a firm, rock-solid handshake.

Occasionally, a problem can overlap both machine- and control-side components. The dealer, who is familiar with both sides of the machine, can determine whether a problem is electrical or mechanical in nature and recommend who is best for the job. When the frustration of getting one side to totally commit to the repair occurs, everybody is brought in and the dealer referees until the fix is finally found. This problem is avoided when machines have a single manufacturer for both machine- and control-side components.

Machine builders are primarily in business to sell new machines. A reputation for good field service is critical to their core business.

1.8   Maintenance Records

After a new machine gets set up, a complete maintenance history is begun. The shops with several machines and experienced full-service maintenance crews keep careful records of all service calls, parts used and any other related service information for a particular machine.

The control-maker, machine-maker and any dealers involved may also keep files of this information in their offices. When a good relationship is maintained with these companies, they may agree to look up the maintenance history of a used machine tool. The machine serial number, machine type, control type and any previous end-user information encourages a speedy search.

Larger operations create and manage a central maintenance library of material. Everything is filed in one place so that nothing gets lost. Small shops tend to stuff the important machine documents into the back door of the machine cabinets. A shop computer can best organize and file such records.

Whenever a new problem occurs, research is done to determine if the problem has occurred before. If previously solved, a very speedy repair follows. Spending incredible sums of money to resolve a previously known problem is clearly better spent on another, more value-added incentive (i.e., take the crew to a machine show).

1.9   Spare Parts

Some shops have spare $900 transistors scattered around like so many old chicken bones. Better this stuff is gathered together and inventoried. There is no sense paying, and then waiting for something already there. Over the years, the odds and ends accumulate; now at today's replacement prices these cartons of miscellaneous elements are a gold mine. Every new machine was delivered with three or four quick blow, high-current fuses. And each can easily cost $100.

Purchasing spare maintenance parts benefits larger operations. The tricky part is deciding what is worth putting on the shelf and what to order as needed. An MBA can do an analysis, collect cycle time studies, compile an *MTBF* (Mean Time Between Failure) and review the service histories, then match the results with the parts available from the vendors. Investment versus downtime is a decision every shop must make.[3]

A shop's previous experience with down time influences a decision to stock a motor or some other large part. Stock is carried to eliminate the serious bottlenecks. Parts for the newer machines are available. Owners of older machines should heed a friendly warning and obtain firm price and delivery quotations for expected replacement parts.

1.10   Machine Warranty

Due to the quality of today's machine tools, the average machine warranty expires without being used. The control

---

builder and machine builder both aim for this result when they calculate the terms of a machine warranty. In many cases, a new machine receives a full two-year control-side plus a one-year machine-side warranty.

The OEMs may also offer extended warranty or preventative maintenance contracts for older machines for a fixed annual charge. Warranty contracts are written two ways: one for parts only, and the other for full parts and labor coverage. Exemptions apply, so the factory should supply a faxed quotation of all the relevant terms and conditions.

Both new and extended factory warranties have one notable exception, the events attributed to *Acts of God* are not covered. The lightning strikes, power surges, earthquakes, rain damage and tornado damage fall outside the warranty terms. A commercial insurance policy, however, may have provisions for these type of losses. Warranty coverage is honorably extended if the present symptoms do not conclusively suggest a surge or strike.

Once a warranty contract is in force, it becomes subject to renewal. A warranty is voided by moving a machine to another company or to an unacceptable machine environment.

Warranty may guarantee a CNC repair, but can remain vague concerning how quickly the work is to be done. For a warranty job, establish a timeline for fixing the machine. On the first day of the problem get everyone involved, put new problems on record with the dealer and builders. Provide as much data concerning the problem as possible. A free overnight shipment of parts fixes most problems the next day. Ask to be placed on the pending service schedule if it looks like a tough problem.

1.11   Telephone Service

The factory service centers have heard it all! On the phone, people talk too fast, too slow, too loud or too quiet. Some calls last six hours straight! Some end abruptly with a few choice words. A week spent covering the phones at a busy factory service center reviews the gamut of maintenance troubles found in the field.

When the phone rings, a few preliminary questions start a phone contact. Be prepared to name off the control-type, machine-type, end-user and warranty status. A machine serial number is used to verify the record. When this introductory information is readily available a service call can be routed directly on to the engineers taking calls. A brief description of the problem gets the engineer up to speed.

After a new caller is identified, a written or computer generated record is created showing the time and date the phone contact began. Making personal introductions and noting any related service number or locator avoids wasting time repeating the same information to someone else in the office.

As the service story unfolds the engineer may ask to have a few things checked; everything is best written down and read back to reduce the confusion. Detailed service procedures are sent by fax to cut down on misunderstandings. Phone queries are taken out to the machine shop and answered before calling back to continue the service investigation.

To avoid someone getting shocked, pinched or smashed by the machine, always question what is being checked and use common sense. Attention to detail is required! If unsure what to do, everything STOPS and the engineer is called right back. The risk is best reduced to hands-off observation. The best phone service avoids disturbing anything until all the important observations have been witnessed. All the screwdrivers and test equipment are kept out of the area; formulating a plan to cure the trouble comes later.

Free telephone service with a talented factory service representative is the best bargain in machine tool service. The good ones are pleasant to talk with, come to the point quickly, and will find the problem. They also follow-up with the customer to check that everything is running fine. In today's increasingly competitive market, customers can come to expect nothing less.

1.12   Factory Field Service

Selling new machines and products is the OEM's mission. The field service divisions were devised to help achieve sales and maintain market shares. Factories could save large sums of money without the field service divisions, but nobody will buy without a viable service option. While building a quality machine avoids problems in the field, an on-site demonstration of complete service ability by a factory's own service team drives up customer confidence and sales.

A factory service visit is set up and scheduled on customer request. Reaction time depends on the current demands for service in the territory involved. Naturally when everyone starts screaming at once it takes longer. Sometimes a new machine or a big customer receives a priority, otherwise it's first come, first served. As the factory parts and people become available, the calls are set up and finished. The standard turnaround is a few days.

Typical charges for non-warranty service calls include travel expenses, hourly labor charges, plus any parts needed. Estimate around $1,500 plus parts, for an in-and-out, one day job. The best ball park estimate is based on the type of problems described and is often given over the phone beforehand.

One advantage of factory service is that they have all the parts to try. Armed with complete documentation, test equipment and unlimited access to fellow engineers, the problem is attacked and finally completed. The toughest calls usually end up on the factory service schedules.

A new machine may receive a courtesy visit to see how things are going and if everybody is happy. Frequently, a modification in software or hardware is installed to update the machine to the latest standards. Any troubles identified at the customer's site are quickly resolved by the dealer, machine builder, control builder or whoever else is held responsible.

The majority of on-site service (both warranty and non-warranty) is handled by dealers, independents or the customer's own in-house maintenance. Factory service calls occur after several people have worked on a problem and failed to put the machine in order. Perhaps a week has gone by, six different parts have been tried by three or four different people, and the job remains to be finished. The factory steps in and may initially suffer a similar fate, but they stay involved to the end because their good reputation (and core business) is now at stake.

Machine shops stick with the service that works—past service experiences explain their new service preferences.

1.13   Independent Service Company

There is a remarkable number of *independents* working in the United States. One-man operations often stake out a city while the bigger operations cover entire regions like Southern California or Central Michigan.

People lured by the freedom of being their own boss set up independent service companies. They "hang out a shingle" and give it all they have. The bigger companies use the less nostalgic "return on investment" strategies.

The independents are on a very personal level with their customers and most rely on word-of-mouth referrals from an assortment of close-knit end-users. Bigger profits arrive for them when established machine dealers aware of a good thing at a good price start sending over repeat business.

Many successful independents gained their experience by working for an established machine tool dealer or OEM, fertile ground for learning the machine tool business. There is no quick substitute for the many late nights spent with the customer, coaxing answers from a misbehaving machine.

Independent, jack-of-all-trades companies have found an attractive niche in the machine tool market.

An independent serviceman in the Southwest region named "Ziggy" earns a six-figure salary fixing machines. Mr. Six-Figures clearly knows the machine tool business. His policy is simple: if the machine isn't fixed, no charges are billed. Attention to detail is imperative; he leaves nothing to chance and wisely uses the maintenance network to guarantee his customers a good, safe, reliable result.

He has wide-ranging skills that apply to several different machine types. He has the ability to start and finish a service call regardless of either machine-side or control-side problems. Many times, he calls around the network from a customer's shop. After talking on the phone, his strong reputation enlists fellow engineers in his cause, with quick and smooth co-operation.

His reputation provides more demand than he is able to help. His new 4X4 is crammed with the best tools and test equipment to satisfy the demand for good independent machine service in his city.

The better financed, large independent service companies offer everything from machine retrofitting and integrated CAD/CAM software to application training and data communication networks. Offerings include in-house software demonstrations, high-speed machining, and the infamous *Internet* presence for a price packages. Others guide in the application of touch probes and parametric programming[4] for a shop's existing machinery.

Some outside vendors offer on-site CNC-tool check-ups. The goal is qualifying a machine's accuracy and repeatability by probing test movements with a laptop-controlled *DBB* (Double Ball Bar) hook-up.[5] The collected data precisely documents the present condition of every machine. More comprehensive machine inspection data is offered by portable systems that measure the reflections of a test laser light source.

4 Parametric Programming for CNC Machine Tools and Touch Probes, Mike Lynch, Society of Manufacturing Engineers, 1997.
5 Accuracy Inspection of NC Machine Tools by Double Ball Bar Method, Y. Kakino, Y. Ihara, A. Shinohara, Hanser Gardner Pubns, 1993.

Another group of companies offer computer communications. They provide fancy communication hubs that allow each CNC machine to talk with the main office or to each other. Yet another group of "clean air" companies offer to stop by monthly to replace or clean every ventilation filter and heat exchanger in the joint.

With so many machines now in the market, independent service companies have bright futures. The independents accept every chance to learn about machines and are constantly increasing their collection of service experiences. Some are at or above the level of the OEM's own employees.

1.14   On Site Training

Gordy

Learning about CNC machine tools is a spotty process; opportunities are often general and hard to apply. The easiest way to get training on a machine is by cornering a visiting expert. The true experts seize the day and show everything they reasonably can. Hanging around and observing a tough problem being solved is effective training.

Gordy says: Make it "show business" with service charges being the admission price. Point out a suspected conclusion, test it using a certain check-pin found in the schematics. Get the customer interested! Carefully install an oscilloscope to monitor the data. While waiting together for the suspected problem to surface, make a small modification to the test program. When everything comes together, the fault surfaces and Wham! The combination of machine knowledge, CNC diagnostics and good problem solving has found the gremlin that has caused so many headaches. This is the training that sticks with a student. Simultaneously, a machine was returned "up-and-running."

On-site training teaches students on the same machines they use everyday. This style of training reviews and establishes good general maintenance policy. The expanded table of contents (Appendix three) offers a list of possible topics for a training class. The network can react to specific training requests.

1.15   Machine Tool Shows

Companies of every kind are selling their wares at the machine shows. The latest and greatest tools are on display with exceptionally qualified people answering questions and trading business cards. Sales representatives are at every turn eager to chat, ready for some brisk business. The machine tool shows gather everyone together in one place. A special opportunity exists to catch up on the latest advancements, talk to people and ask a whole lot of questions. Interesting technical seminars are often presented, and down in the back corner of the basement level, all the newest books are available.

Behind the scenes, the OEMs have made expensive preparations in anticipation of any possible maintenance problems. To showcase a down machine at the show is the worst possible scenario. Factory spare parts, documents and people are all connected by a high-tech rapid response cell phone and beeper network.

Gordy recalls some interesting events that occur at the shows. First off, the models and professional cheerleaders are always a big hit with the machine tool crowd. Also many lively stories from the road are relived face-to-face among the many business friendships established anonymously over the phone. Eager machine designers have been known to jump up on a competitor's machine and start inspecting the machine casting and constructions.

Salesman

Accidents can also happen. The plush carpet laid out at one booth created static shocks, zapping one new machine out of operation. Another machine was driven full speed into a fixture vice, turning a lot of heads and causing a stampede to see the mechanical carnage. Everyone asking pointed questions, dealt terminal embarrassment to the red-faced operator.

Hospitality is secretly extended to all the dignitaries in well-stocked hotel suites. The food, liquor and fun is searched out and found by many a show-goer.

Big shows in the U.S. include the International Manufacturing Technology Show (IMTS) at McCormack Place in Chicago, held every other year in September, and the WESTEC show in the Los Angeles during March. Both Europe and Japan[6] have machine shows of international importance.

Many smaller machine shows and open house-type events are given at the local level. These are usually promoted

---

6   An international show called the Japan International Machine Tool Fair
    (JIMTOF) is held once every two years, about a month after IMTS.

in trade magazines, with invitations sent to all sponsors, good customers and business friends.

1.16   Complaints

Creative complaints are really an overlooked art form. Some complainers are so effective at getting results that meeting them in person almost becomes a respected event. Of course, anyone with a bunch of new machines can skip this section. For them, one blunt call does it all.

A factual, chronological listing of what is happening written into a clearly worded statement distributed by letter, fax or phone to everyone in a position to do something will find results. Effective complaints have a subject and a target. A valid complaint directed at the wrong person will not accomplish much.

The best policy is one that is pleasant and firm with numerous follow-ups and timely reminders of specific, constructive actions that could solve the problem. Once commitments are received, everything should be relentlessly reconfirmed to keep it on the front burner. Overly aggressive methods win short-term battles, but can backfire in expensive ways down the road.

If getting upset is the only way to capture a company's attention, then it is time for a change. Letters containing a valid complaint will get noticed in today's customer-driven service environment.

Working with the network gives great results every day. Sending letters that praise an individual's actions cost the same and go a long way in establishing an appreciation that consistently gets things done.

CNC QuizBox

1.5 Where in this book is the information for:
   a.) Specialized machine documents?
   b.) Factory published documents?
   c.) Training outline?

1.16 Write an effective complaint letter and an effective letter of praise.

2 *Safety*

2.1 Introduction

Take every precaution while working around machinery. Build work habits that eliminate the possibility of personal injury, not just reducing or minimizing injury, but eliminating the potential chance of injury. A machine with enough power to cut through metal can seriously injure or kill the human operator. The scope of the dangers are briefly illustrated.

Mechanical dangers. If the walls of a machine shop show evidence of flying chuck jaws, or if holes are found in the windows and doors of old machines, danger has paid a visit. If he's still around—find the guy standing in front of the machine to hear how it happened. Injuries happen quickly. Accidentally dropping a sharp, heavy tool can sever a finger or break a toe on the way down.

Electrical dangers. A spark from a single misplaced voltmeter probe can leave an entire shop in the dark. The stunned technician is left wearing black soot to the wrist, thinking of a career change and why on earth the power was left on.

Environmental dangers. An unexpected slip in a pool of cutting oil gives a nasty fall. Oops! Head first into a live open cabinet, the head and right arm bumps across 220V terminals, instantly flashing off the skin and hair—three unforgiving errors that can never be taken back.

As a rule, people in machine shops are safety-conscious. They have seen, heard about or had a close call with accidents serious enough to forever change a person's life. (That's more than enough experience to drive home the point of shop safety.)

As a result of work place danger, new machines include elaborate safety interlock systems and legal disclaimers. For years, free advice for fixing and checking machines has traveled over the phone to grateful CNC users. This policy is based on the assumption that everyone understands the danger and applies utmost safety while working around machine tools.

2.2   Electrical Safety

Mistakes leading to personal injury also cost thousands of dollars in electronic destruction. Safety protects the machine and the person. When rushing through a repair to save time, a single mistake destroys everything in one fell swoop. It happens that quickly.

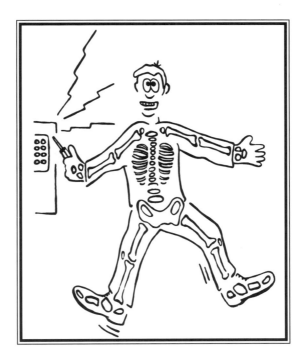

Dangers

Before opening cabinets, **power to the machine must be turned off and locked out** to reduce the risk of getting shocked. Also, **test equipment should not be installed to live circuits**. And **when the power is re-applied, everyone should stand clear**. These three tips would fit nicely on the back of every voltmeter in the field.

Hidden capacitor circuits remain charged long after the power is turned off. As on old TV sets, these large caps are often found inside the motor inverters of CNC machines.

Shorting a cap is dangerous and explosive, due to its basic electrical nature. (An exercise in the QuizBox asks for the technical reason supporting this statement.)

Qualified engineers avoid surprises. Their mental and physical moves are judged for safety in advance. All loose metallic objects are kept away from the machine. Metal clipboards, wrist watches, jewelry and loose tools that cause shorts are kept out of harm's way. Periodically, the service effort stops, allowing time to straighten up the area. Electronics are best left to qualified people who understand the *hidden* dangers.

2.3   Mechanical Safety

Heavy mechanical work should be left to qualified people who know the hidden mechanical dangers. A visiting control-side engineer instinctively calls in a machine builder when mechanical replacements become necessary. A mistake while replacing a motor linkage allows the weight of a large machine casting to come smashing down. Any fingers caught among the freewheeling gears, belts or screws may come up missing.

When mechanical things are moving, hands, hair and clothing can get snatched. Time is well spent identifying all dangerous possibilities because one mistake in operation can send objects flying out of the machine at ballistic speed. The CNC-operators have the specialized knowledge needed to run these machines—their cooperation is a vital part of a safe repair.

During a service call operators are reminded that they are in charge and that everyone's counting on them to operate the machine professionally and safely. They are accustomed to using lightning-fast fingers during everyday running, they need to slow down during a service investigation. A testing program may reveal some unforeseen problems. The point is made direct and clear—don't rush things during a service. There's no need to impress anybody, just override the machine speeds and keep the sliding safety doors shut. If something looks wrong—STOP!

> Gordy, our seasoned service engineer, was absolutely shocked! This operator was talking endlessly about his illustrious experiences while flying through a new test program. The purpose of the test was to check the machine's positioning accuracy, when suddenly, the machining center picked up an unexpected 8000 rpm spindle command. The tenths indicator mounted in the spindle hung on, up until about 5000 rpm before exploding. The pieces slammed into the quarter inch Lexan shields leaving a bullet hole result, directly in front of Gordy!

A service investigation creates an unusual time and circumstance in a machine shop. These special circumstances inevitably increase the chance for something tragic to occur. Any confusion or distractions in the shop must immediately stop a service investigation until a slow and methodical operation is completely understood and accepted. No one needs to be injured during a service call.

Mysterious machine problems result in accidents. Such situations recommend a complete and immediate shut-down of the machine until the problem is identified, understood and rectified by the OEM. This may be an unpopular position but concern for the shop workers is much more important.

The first rule for new service engineers is that if they feel unsafe about anything, they need to get away, get out, go to a phone and call the boss. Without exception, unsafe conditions stop a repair. Safety problems are assigned a higher priority than maintenance problems.

2.4   Safe Environment

Machine shops contain metal chips, cutting fluids, oil-smoke and lube-oils. This is the normal environment for NC machining, and automatically causes concerns for safety.

Simple common sense safety measures are in order. Before starting work, the floor around the machine is given a good sweep. Some cardboard or kitty litter can soak up or contain the stray oil. A sturdy chair and bright, plastic, clip-on fluorescent lights are also helpful for lengthy investigations.

Anyone caught without safety glasses in a machining environment needs a good chewing out. Safety glasses with full side-protection and clean clear lenses are a must-have. A few extra pair should be kept at the ready for any unexpected visitors.

> During a service call, Gordy unknowingly brushed up against a water spigot. The water began seeping under and around the machine. When he noticed all the water on the floor, he left the customer and called his boss. While the customer absolutely wanted him to keep working, the service was temporarily canceled until the water hazard dried up.

A nice comfortable pair of steel toe safety shoes help protect the feet and toes. Eardrums can be protected from the numbing din of machining noises by a set of expandable foam ear plugs. A disposable oil-proof suit can save clothing and avoid pesky skin problems on bigger jobs. Protect the hands during a greasy replacement with rubber dishwashing gloves.

All these items need to be kept handy for the hard-working shops out there. Wearing every single piece of protection equipment—including a gas mask—sends a message for better periodic cleaning. Some shops have trouble getting service because of their overly toxic environments.

CNC QuizBox

2.2 What is the relationship between voltage and current across the terminals of a capacitor? What is the implication for safety?

2.4 What is the resistance of a human body? With the feet grounded in a puddle of water, what is the current through the body for a 200 Vac shock?

3 Computer Numerical Control

3.1 Introduction

Computer controlled machine tools are expensive, complicated systems. Everyone who looks at a new CNC machine sees something different. Salesmen hawk the 1500 inch-per-minute rapid speeds, the 32-bit computers and the two second chip-to-chip cycle times. The computer engineers notice the open bus architecture, I/O strobe rates and application sequence programming. While the service engineers generally agree these are all interesting topics, they seldom apply in fixing an ailing machine.

The flowery technical discussions apply to designing, building and, perhaps, selling the new machines. Once the design is finished and the machine is sold, there are no more changes. Service aims to understand everything about the existing system from the factory. Then, when a problem develops, just put the machine back within the original factory condition, and the problem will be solved.

In the last twenty years, the machine systems have consistently improved. The generations of older technology have led to hundreds of design modifications and improvements. Computers, motors and machines are now much faster. Interestingly enough, while much has changed, a common bond or legacy survives to evaluate all the different CNC machines.

3.2 History

Old manual engine lathes and knee mills had no computer. The operator cut good parts by cranking on big shiny hand wheels, expertly coaxing the tool down the desired path. After years of this back and forth dialing, the crank handles became hand polished to a mirror finish.

The idea of using electric motors to do all this cranking came up during the trouble-free 1950s. The big question was whether a motor could crank as accurately as a human. An interesting solution written back in the '50s was recently posted in the hallways of a famous engineering university. Using the dual premise of electronic control and motion feedback, a written proposal to the cranking dilemma had long existed in academia (and the military establishment).[1] Apparently, these designs didn't move into widespread commercial manufacturing.

Most of the early motorized machine tools were temperamental, expensive and hard to set up and operate. Early on, the market potential for dependable CNC machine tools was solid. During the middle and late 1970s some exceptional computer numerically controlled machine tools were introduced to the world market.

The factory standards used in the design and manufacture of these early CNC machines routinely provided years of flawless operation. A single machine in the late '70s gave the profitability to purchase one new machine after another. Shops with a few of these mighty machines have done well to the present day.

The basic design of the systems which were mass marketed in the '70s bore a resemblance to the posted designs from the '50s. Accurate control of electric motors was accomplished by closing the motion loop using early micro-processors and digital circuits to track the command pulses going out and the motion feedback pulses coming back.

First generation machines naturally use older technology than what is sold today. Back then, the motor current for axis and spindle motion was provided by thyrister-controlled, *DC* (Direct Current) motor drivers. Internal computer logic and control functions came from an 8-bit *CPU* (Central Processing Unit) built on a platform of intricate, pressed wire-trace circuit boards, and all were plugged into a collector's item wire-wound back plane. Memory data originally resided in old

---

1   An interesting treatment of the history of CNC is given in the August 1996 edition of American Machinist.

magnetic-donut memory boards, which has since been replaced by compatible, modern *IC RAM* (Integrated Circuit–Random Access Memory) memories.

The old 8-bit computers sense machine movement by counting signals from the precision *feedback* units mounted to each motor shaft. These units send motor direction, speed and positioning information. Although costly, the complete feedback units were designed to be replaced or exchanged.

Early CNC turning centers could move at 300 *IPM* (Inch Per Minute) and hold two-*tenths* of an inch. The smallest position unit the computer displays is 1 um (One Micron, Metric) or 1 tenth (One Ten-Thousandths, Inch). An active change of the machine position flashes over a cluster of universal seven-segment displays. In concert, the displays form a somewhat coherent read-out for the operator to use in programming and monitoring the machine.

Machines from the seventies are still running in production today—some 20 years later! The best explanation for this longevity is a combination of stout, oversize mechanicals and top-quality wiring and electronics running at the typically slower speeds.

The new machines of today are fast. Performance AC motors are driven at high speed by compact vector-controlled digital inverters. But more astounding are the computer improvements: the new high speed *RISC* (Reduced Instruction Set Computing) processors, and the running of expanded software options from open PC platforms.

Machine functions previously handled mechanically are now economically accomplished using nifty NC software models. Industrial grade PC computers are now upgraded with plug-in NC modules. Similar to the process of adding a modem or installing a new software program, an entire numerical control is plugged in and configured for a machine.[2] The closed CNC systems of the past are now beginning to open up.

The twenty-year transition from the first generation machines to the systems of today frequently visited the concept of *Kiazen*, the study and implementation of continuous process improvements.

3.3   Common Features

A generous enamel finish protects the surfaces of a CNC machine tool. Close-fitting metal covers and cabinets provide a pleasing shape, while safely concealing the moving parts within. Direct views of the metal cutting is provided through clear observation windows sealed into large sliding safety doors.

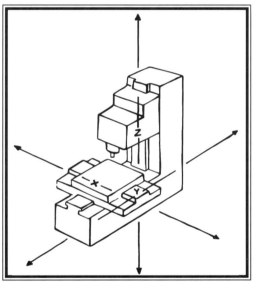

Figure 3.1 Simple CNC Machine

The operator's panel is inviting and friendly, with clearly marked knobs and switches all strategically located. Behind the cosmetic covers, a regiment of hidden motors, junction

2   An interesting article on converting an old CNC to a PC is given by Mark Albert, "Software is the CNC," Modern Machine Shop, November 1997.

47

boxes and cables are found. All smartly placed and factory-sealed against the ingress of swarf (shop debris) and oil.

Around the back of a machine are the locking control cabinets built of heavy gauge steel, firmly bolted to the underlying machine casting. Tight-fitting rubber door seals protect the sensitive CNC electronics inside from the expected levels of dirt and oil in a machine shop.

CNC
Designer

Inside the cabinets is a careful layout of control electronics. Nothing is left to chance here; the secret of the quality can be seen in detail. All the wiring and cables are neatly tucked, tied and screwed into place. Each switch, fuse, wire and terminal is clearly labeled and cross-referenced with factory supplied cabinet schematics. Heavy, locking connectors and terminal strips are used throughout.

Inside the control cabinet is also a box-like rack called the CPU rack. This rack holds all the separate *PC-Boards* (Printed

Circuit Boards) of the computer's brain. The CPU for a machine tool is generally a collection of circuit boards.[3]

With the proper OEM-supplied procedures, replaceable boards quickly plug in and out of the CPU rack. Parallel bus connectors, at the back of the rack, allow communication between neighboring boards. Found next to the CPU rack are the servo-units, power supplies, spindle drives and an assortment of machine relays and magnetic contactors.

The electrical parts buried inside a CNC machine are inter-connected by literally miles of wiring. Fortunately, the factory devised a clever wiring system to simplify the task of repairing machines in the field. A provision for replacing bad parts, or for locating some obscure control signal, was considered by the manufacturer. A full appreciation of the factory wiring scheme only comes with experience.

Larger bundles of wire are terminated using one of two common styles. In the first style, wires from a bundle are individually separated wire-by-wire and screwed down on long strips of labeled *screw* terminals. Single wires with labeled sleeves then run from these strips out to the correct components.

The second style of wiring uses compact, multi-pin plastic connectors that quickly *plug* into all the PC-Boards and nearby units. Plugs running vital control signals are plugged in and safely screwed into place. These miniature plugs are designed and built by factory vendors.

To summarize the plug production process, a raw cable has every wire stripped and crimped to miniature metal pins. These pins are then carefully inserted into a specifically numbered slot on a connector plug. A gender is assigned to help keep the plugs and pins straight (for example, a female plug

---

3   Appendix Two lists common questions the OEMs are asked daily. Questions concerning board functions, replacement procedures and factory recommended testing procedures are outlined in this listing.

with male pins is considered a male connector). Each plug is given a name, and each single wire within that plug is given its own distinctive pin number and signal name. Incredibly, all the wire labels, plug names, pin numbers and component stickers actually match up with the cabinet schematics sent with the machine.

3.4 Up and Running

A qualified operator gets the machine "up and running" by flipping on the main breaker and pushing the "on" button (sometimes twice). Driving around a $300,000 machine is a big responsibility; only a qualified operator who is willing to take on this responsibility can give the proper guidance during service. The operator may decide to go slow, keeping one hand on the red emergency-stop push button.

With the machine humming at full servo power, it is ready for an axis zero-return. Pressing the "automatic return" push button causes the machine to start moving. Each controllable axis moves rapidly toward its homing position known as zero return (ZRTN). The machine suddenly slows and, after a moment, the zero return LEDs for each axis start blinking on. Everything is almost ready for the machine's primary purpose–cutting parts automatically from a standard CNC part program.

Final preparations for production are the manual operations that respond directly to the operator's commands. From the operator's panel, a tool position is set by turning a rotary switch and pushing to the "index start" button. The entire turrent, containing all the tools, rotates into a new position. Spindle speeds are commanded by a speed rheostat knob and "spindle start" push button.

With all the zero return LEDs lit up, the machine can run single automatic operations in *MDI* (Manual Data Input) mode. In MDI mode, a single axis move or spindle command is requested by keying in the command data and executed with a push of the "Cycle Start" push-button switch.

When a machine is running in production a part program in memory runs over and over, relentlessly producing the small mail sorting wheels ordered by the post office. After each 20-second cutting cycle, a brief five-second pause is ended by the drop of another finished part. The parts roll out the front of the machine and onto a slowly moving conveyer belt. During the machining pause, a new *blank* piece of metal is quickly picked up and placed in the machine by a small speedy robot with only one arm. After the arm clears, the machine snaps back into action and buzzes out another finished part. This production cycle continues, twenty-four hours a day, seven days a week, without shutdown. The same part may run non-stop for years.

3.5   Control Hardware

Avid computer people understand the distinction between hardware and software. *Hardware* describes all the hard electronic components, wires and circuit boards connected together to build a *platform* for running the application software packages.

All modern CNC machines have hardware and software. Older machines use a higher proportion of hardware to software because they were designed in the days when hard analog circuits performed many of the tasks needed for motion analysis and control. Newer machines have systematically replaced this old analog circuitry with more efficient, software-controlled, digital circuits.

Common pieces of hardware include the main CPU, the motion CPUs, keyboard interface and memory storage. Newer machines can also include extra hardware platforms for running color graphic CAD/CAM software.

The CNC memory space is expanded using add-on or upgraded memory boards. The option of memory expansion is expensive on the older machines running proprietary control platforms. Aftermarket companies offer memory upgrades for some of the older proprietary controls. An alternative to more hardware memory is *spoon feeding* large data files with standard desktop computers or upgrading the outdated controls with new open architecture systems.

Remote pieces of the hardware are linked together with high volume *serial data* lines. For example, the operator display at the front of the machine is driven by streams of serial-data sent from the computer rack in the back. Still another stream of serial-data relays the ongoing status of every panel and keyboard push-key. Cables are run throughout the machine to carry these vital serial-data streams.

Serial data streams are handy for moving a lot of data on very few wires. The transfer of CNC programs over an RS232 cable is a good example of serial-data transfer. During transfer, the data stream, or individual *data packets* are very hard to recognize–the important thing is receiving and decoding an exact copy of what was sent, every time. Elaborate communication protocols accomplish this job.[4]

The NC computer efficiently controls the entire machine by reading and sending serial data. The CPU understands the serial-data received from the outside world and sends along the proper command response. The machine then dutifully follows the computer's direction and relays additional serial data. In the case of a panel keyboard, each and every key switch contact is *strobed* into a *real-time* status and then converted into a serial-data stream.

Rows and rows of switches and light bulbs that are mounted out on the machine panel are merely "Inputs and Outputs" to the mighty computer. Computer addresses are assigned to some of these *I/O-Signals* (Input/Output Signals). If the computer senses a *switch input* address change, it may change the *lamp output* address to light up the pressed switch.

---

4 See also Section 5.5.1 and Section 7.9

To an operator, the light came on above the switch, but the computer *may* actually have sent this bulb response.

Combining the ideas of the I/O-Signals with serial-data transfer leads to the discussion of the most frequently replaced piece of CNC hardware—the infamous I/O-Board, or more accurately, the PC-Board handling the serial distribution of the many machine input and output signals.

Now, for a less technical picture of how the CNC connects to the outside world. Imagine riding high overhead in a helicopter. Out the window is a congested, two-way highway with **eighty** lanes of traffic traveling in both directions. Suddenly up ahead, the highway reduces to a single two-lane tunnel. The north-bound *Input* traffic is riding calmly along in their own lanes, never changing into someone else's lane, until right before the tunnel where everybody miraculously shoots into one north bound tunnel lane without a scratch. After the tunnel, the inputs all branch back out to the same lane as before. Of course, the south-bound *Output* traffic is driving a similar course, but in the opposite direction. Naturally, they all drive really, really fast (it's only an analogy).

In this analogy, the tunnel traffic represents a serial data stream, and the miraculous merging and un-merging at the tunnel entrances describe the function of the I/O-Board circuits. This analogy considers each lane a channel and each car a signal. Newer CNC machines just support more lanes on the highway. If anyone breaks down on the highway, the I/O-Board generally receives the blame.

The I/O-Board is the interface between the deep internal world of the computer and the recognizable world of the machine. Later, powerful diagnostics are identified to test each and every channel of the I/O-Board.

The replacement of I/O-Boards and other pieces of computer hardware is routine. In each generation of control technology are PC-Boards having a distinct model type and manufacturer. Getting the correct written replacement procedure from the factory-authorized service centers starts with correctly reading the board's model type.

Verbal procedures are not as reliable as written *exchange procedures* used successfully by others in the past. Before a service engineer pulls out a board, a call to the network is made to get a good faxed procedure. When procedures are too complicated or just plain unavailable, an on-site OEM service is scheduled.

Before hardware is exchanged, the correct replacement parts are on hand and the factory's exchange procedure well understood. With the power off, a rough picture showing the original orientation of the system being changed is sketched by the service engineer. Wires, screws, connector labels and settings are in the drawing. Crossed wires cause heavy losses and cross bosses.

In general, the new board will look like the original. The original and new boards are placed side-by-side on a good clean spot for comparison. All the jumpers, dip switches and socket chips are checked. To preserve the option of going back to original condition later in the repair all the board serial numbers are recorded before anything gets mixed up. If something doesn't match up, or doesn't look right, the factory is called with all the specifics before proceeding.

Some exchange procedures require saving the *control data* before starting a replacement, this means the end-users' personal data (like part programs, parameter settings and tool offsets) are saved by computer before the work is started. Whenever the machine is shut down, some control data is kept alive in non-volatile hardware memory components.

It is good practice to save the NC control data and verify its content before a new job is started. With this data saved a dead machine can be brought back to life. Most shops routinely save and verify the control data from every machine in the shop.

3.6   Control Data

A sudden power surge, lightning strike or communication foul-up can cause the control data to accidentally erase or scramble. Remember, the control data (or NC-Data) is the valuable information, like the part programs, parameter settings and tool offsets stored inside the NC computer. Occasionally, just pressing the wrong set of keys can cause a heart-breaking scramble.

To recover from such an event requires that a copy of the NC-Data was saved on disk or printed out on a hard copy before the scramble occurred. Lost data translates into a few days to a few weeks of rebuilding from scratch. Backing up the NC-Data permits a quicker recovery.

Of all the types of control data the part programs written for the machine are the most widely understood. Full-time operators and programmers earn their living commanding the CNC with a mostly standard programming language called *G-Code*. Visiting service people are notoriously lousy programmers in comparison with the shop professionals or other full-time applications people.

A basic understanding of programming is helpful during NC maintenance. Experienced service engineers understand it is best to let the operators do the operating while the service people concentrate on the repair. The operators who live with the machine daily can quickly set up and write a short test program that can effectively evaluate the machine.

To make a simple program, operators follow a list of program steps that are certainly worth mentioning. First, a positioning system for the program is chosen and tools are selected and given offset instructions. The actual cutting path the tool follows is calculated and added into the program. Choices for material and tooling determine the appropriate cutting rates, or feeds, for each tool called in the program.

Finally, the preparatory M, S and T codes are selected and placed into the G-Code program. The machine function M-codes include: coolant spray, door closed, waiting tools and rotation forward. The speed function S-codes specify spindle speeds, and the tool function T-codes pick up the tool change and offset instructions.

The operator carefully verifies a newly created program by closing the safety door and running the program in *single block* with all the machine speeds reduced. The first cycle performs *air cutting*, where the tool path is safely shifted away from the true path. A five- or ten-inch position shift is keyed in manually by the operator. When everything is thoroughly checked and approved by the operator, a metal blank is clamped in the machine. Slowly at first, then faster, the cutting chips begin to fly.

A basic ingredient in the recipe for making money on CNC machines is part programming. An operator develops impressive, all-around programming and application skills by setting up and running new programs daily. Textbooks and classes on CNC programming are widely available.[5]

Another type of control data is the collection of internal machine software settings grouped into the machines *parameter list*. The majority of these settings are found listed in the back of the operator's manual for the machine. A brief and sometimes cryptic description is written for each user-accessible setting address.

Changing an unknown parameter is very dangerous. Before any control data or parameter setting is changed the originals are carefully recorded. Next, the meaning is well understood **before** making changes; accidentally changing an unknown address can permanently lock-up the machine, cause

---

5   Classes are offered through professional trade organizations like SME, Society of Manufacturing Engineers. Machine builders and dealers also offer classes for the systems they sell. Programming classes on video tape are available.

a runaway, or worse. A parameter change is always tested against the expected result.

The secret factory parameters are not listed in the operator's manual. Any changes require factory analysis and approval. Control builders use exotic procedures and boot-up files to safeguard these hidden factory parameters. The factory knows the distinctions between user-level and factory-level settings. In either case, without knowing its meaning, a parameter is never changed.

Some parameters toggle like a switch and others specify a value. A *toggle* parameter switches a desired function ON and OFF (1=ON, 0=OFF). A *value* parameter carries a numerical value like twenty seconds, chosen anywhere within an allowable range of defined limits. Both types are keyed into memory from the operator's keyboard. The procedure for changing parameters is given in the operator's manual.

Once all the parameters are correctly set, the machine performs the way the customer desires. The speed of a machine or the number of tools in the tool turrent are both supplied in the parameter settings. There are so many functions on the new machines that thousands of parameter addresses are assigned—the old machines have less than a hundred!

At the factory, a basic set-up list is calculated and transferred into the new machines. After an on-site installation, the basic parameter list is adjusted to the preferences of the customer. Any changes from the standard set-up list are penciled in on the original factory copy and then stuffed back in with the original machine documents.

If the parameter memory is electronically scrambled after a power surge or lightning strike, the machine quits until a data re-boot, or regeneration, is completed using the correct parameter list. This process is like cleaning out a dirty closet and getting a fresh start. Dirty, polluted NC-Data is cleared out and the computer system is initialized to receive a fresh new batch.

Accomplishing a *regeneration* procedure is fast and easy when a good list of control data was previously written down or kept on diskette. A copy of the correct regeneration procedure is at the OEM, and can easily be sent over the facsimile machine. Care is taken to avoid accidentally clearing more data than is just absolutely necessary.

Once again, current back-up lists of NC-Data are vital. Good preventative maintenance avoids the hassle of recreating a lost list of parameters and programs.

3.7   Control Software

The instructions for making a CNC *act* like a CNC are written into the control software. Every builder has pet names for this software, or collection of software, usually the terms *NC main software* or control software adequately covers the subject.

A circular move is performed by typing simple G02 commands into the program. The machine relies on hidden routines stored deep in the NC software to understand and execute this standard G-Code command. Every machine understands G02 is a clockwise, circular move. However, a program with more elaborate G-Codes, such as canned repetitive cycles or tool radius cutter compensations, will not run between the different control builders. Because each control builder wrote a slightly different routine in their NC-main software, a program code modification or *filtering* function needs to be performed (a CAD/CAM post processor can accomplish this task).

The control builders are smart, they use proprietary routines to help maintain business arrangements with their machine building partners. Each machine has only the special NC functions the builder purchased or helped initially develop. The operator's manual lists all the possible G-Codes understood by the NC computer. To see how many functions will actually work on the machine, compare the control builder's entire list with the machine builder's selected specification.

When an NC main software problem occurs in the field, it is carefully documented and sent to the factory for engineering approval and modification. The factory investigates and, if required, modifies the internal software code. The newest version of NC software carries all the latest modifications and improvements. A written representation of this type of software exists only in the factory archives.

The first generation of CNC-tools installed the NC software data using long spools of punched paper tapes. Today, just a set of plug-in chips or software data files are replaced when the NC main software is upgraded to a newer factory version. Like most computer software packages, only the most recent version is shipped from the factory.

After a few years the updates stop coming. By then, all the major bugs have been reported and worked out of the system. A careful listing of software updates and improvements are kept on file so any newly reported problems can quickly be identified as coming from outdated NC main software.

3.8   Sequence Software

In the machine tool business, several combinations of competing equipment are available. In one shop, three machines can appear identical until a closer inspection reveals three keyboards and operation panels from three distinct control builders. How can three identical machines have different style control computers? The flexibility is provided by the *sequence software* written specifically for each machine/control combination.

Originally, the control and machine were separate, waiting to be wired together. Rather than trying to hard-wire everything together, software mapping was done between all the machine signals and control signals. This software mapping effectively joins the control to the machine. The result was burned into a couple of chips and plugged into the control computer—incredibly more efficient than running wires. By writing a new version of sequence software the builders can offer the same machine with different controls, or the same control on different machines. Powerful stuff!

Machine builders write the sequence software during final design and development of a machine. Sequence software is also known as application software, PC-software, PC-ladder, ladder software, or just ladder. Others use the familiar term *PLC* (Programmable Logic Controller) when making reference to CNC applications or to stand-alone programmable equipment in general.

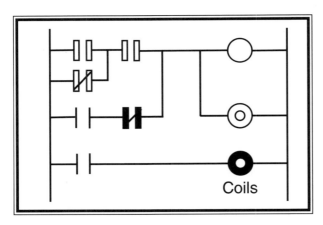

Coils

Ladder

Sequence for a machine is represented in a *ladder diagram*. The ladder mimics the old *RLL* (Relay Ladder Logic) diagrams supplied for relay circuits from the '50s. The sides of the ladder represent "power lines" with the different sequence program elements added to form the rungs. All the sequence software information is "concealed" in this written ladder form. In the right hands a ladder diagram is a powerful maintenance tool. In the wrong hands, the ladder sends people chasing up waterfalls and reporting stray leprechaun sightings.

Learning to read a ladder is easy; understanding what's really going on is the tough part. What usually happens the first time the ladder diagram is opened is a suspect signal address is quickly found. A signal condition here causes this condition over there, unless, of course, this condition over on the next page. . . um . . . which is caused by this other condition exists. What?

The sequence ladder presents a challenge. Some controls electronically display the ladder diagram with time saving search, tag and status functions to make navigating the sequence easier. An exciting new development in this area is the "flow chart" type sequence offered on PC based controls.[6]

Each builder decides their own style of ladder symbols and layout. All the symbols and connected pages are all linked together, so a quick follow-up of any signal status between machine and control is plausible. The condition of most ladder addresses are viewed directly using the CNC internal diagnostics or ladder display screens.

3.9   Motion Control

Help wanted: Machine tool desperately needs reliable computer for obvious assignments. *Supervise machine motion and carry-out desired commands.* If things get out of hand, set off alarms and shut everything down. Plain or fancy welcome to apply. Salary commensurate with capabilities.

This is the job description for NC computers. The key jobs are: (1) motion control; (2) making different style motion commands; and (3) detecting alarms. Every control accomplishes these three objectives. Old and new, plain or fancy, all must carry them out. Some do it with decidedly different levels of sophistication.

High-quality machines deliver motion control for years without a single failure. These systems also offer the flexibility to expand into neighboring applications by simply adjust-

---

6   See the reference given in section 14.14

ing internal control settings and software choices discussed in previous sections.

As a reminder, motion includes position, speed and acceleration. A physics class will introduce the formulas of motion, and explain how inertia, force, mass and acceleration are all related. Just take the weight of a machine, calculate the motor size and put it all together. What's the big deal? Well, beyond accurately solving the motion problem comes finding the electronics capable of supporting the solution.[7]

A *closed-loop* system gives these machines their incredible motion control performance. These "motion loops" are closed by powerful motors, servo-amplifiers, motion feedback signals and the mighty NC computer. For now, the focus is on the role of the computer in closing the motion loop. The role of the other parts will come up in later chapters.

Getting a grasp of motion control is simplified in this book using the *friendly pulse* analogy. Imagine a little pulse traveling around in every motion loop. The pulse changes its disguise as it moves from one end of the loop to the other, but whenever passing near the computer, it is always recognized and counted as one *pulse*. The computer needs these friendly little pulses—without them, there's no motion. With one pulse present in the loop, the machine will move a little bit. When teams of pulses start running around, the machine moves farther, and faster.

These pulses are interesting characters because each one moves the machine the exact same distance. This distance is given in the operator's manual under the name *minimum increment*. The minimum increment is the smallest distance displayed on the computer visual read-out. Commonly, the

---

7 Read more in an article by Chuck Raskin, "The Science of Tuning Servo Motors," Motion Control, September/October 1997. Check the web for sites offering free software based tuning programs available for download.

minimum increment distance is set for 1 um or 1 tenth (0.001 mm or 0.0001 inch).

Thinking in terms of equivalent friendly *pulse* and minimum increment helps simplify the discussion of positioning a modern CNC machine tool.

To start the discussion, make the machine move one pulse. How? By turning the manual hand wheel one click (first ask the operator's permission). This commands a one pulse move, the minimum increment. When the hand wheel clicks, the position read-out advances and the machine moves the smallest distance it can. Handle forward one click and the machine moves forward one pulse. Handle backward one click and the machine moves back one pulse (to zero).

When the handle clicks, the NC computer launches a pulse into the loop (a command pulse). This pulse races around the loop, causing some motion and then returning like a boomerang (as a feedback pulse) to the computer. The computer is happy; it knows full motion was achieved because the friendly pulse came home from its travels.

The computer starts the motion process by sending out a single *command* pulse. This pulse is converted to a *motion* pulse by the servo-motor combination. When the motor rotates the machine's minimum increment distance, an equivalent *motion* pulse is created. This motion creates a single *feedback* pulse which is returned to the computer and finishes the process. Everything in the loop worked beautifully, the one click move from the handle quickly translated to a one pulse, minimum increment shift of the machine.

All the other kinds of machine motion (and there's quite a few) obey this simple pulse analogy. The motion command options, common to all machines, are the three *modes* of travel named: handle, feed and rapid. These modes achieve the *overall functions* of the machine—things like multiple-axis interpolations, or smallest minimum-increment moves. In the newer CNC machines, the options for motion control have built and expanded upon these common themes.

In practice, the computer launches millions and millions of pulses and steadfastly requires a feedback answer pulse for each. Nobody gets lost. The machine only becomes *in-position* after finishing this circular *pulse distribution* dance.

A simple movement, in program, is achieved using either *rapid* or *feed*. Using standard G-Codes, the G00 (G-Zero) commands a rapid move; the G01 (G-One), a feed move. The "rapids" quickly position the machine and the "feeds" are highly controlled rates for metal cutting. The numerical control computer provides these motion modes by altering the "shape" of the loop motion commands.

The *shape* of the command in feed-mode is smooth and exponential, while the quick rapid moves have steep, abrupt linear accelerations and decelerations Other nifty shapes are synthesized by the newest computer controllers.

The benefit for using G01 feeds is motion synchronization, where speed in one direction is exactly related or *synchronized* with speed in another. A feature found on turning centers is called *constant surface cutting* (G96), the synchronization of one axis cutting speed with the spindle rotation speed. The simultaneous, controlled feed of several independent axis motors, or *multiple axis interpolation*, allows linear, circular, or helical cutting movements. The mighty NC computer controls all the motion loops simultaneously, generating appropriate commands and counting feedback in *real time*.

A large machine moving at high speed is commonplace in shops around the world. Imagine if the motion control was suddenly lost! These lumbering CNC giants would be potentially dangerous. To safeguard against this event, sophisticated alarm detection systems are built right into the computer's hardware and software. The alarms trigger a chain reaction of events to safely shut down and stop the motion of everything on the machine.

Critical alarm routines are programmed directly into the control computer. These routines continuously compare dynamic internal computer data against safe, acceptable test limits. If the computer's soft data crosses the limits, an alarm is captured. Again, the *soft* alarms are described and detected in software routines. No hard alarm contacts close, only a brief dynamic error was detected inside the passing computer data.

The captured alarm is displayed in the control alarm pages. A good example of a soft alarm is a servo loop error. If the accumulated lag pulses within a servo-loop exceeds the software maximum, a software positioning error is latched and displayed. This reflects a condition where the machine, for whatever reason, just didn't get where it was supposed to go. (Maybe a weak motor or heavy servo system is the culprit.)

Alarms from hardware are more straightforward. They are triggered by some physical alarm switch contact on the machine. Keep in mind, the NC alarm page may help locate this mystery alarm contact, but the numerical control didn't detect the alarm—it just received the news and displayed it. When the source of the *hard* alarm is found and reset, the NC alarm will also clear. Additional service measures are indicated if an alarm returns after the faulty unit is reset.

CNC QuizBox

3.5 Draw a simple example of a:
a.) Digital circuit
b.) Analog circuit
c.) Mixed analog/digital circuit

3.6 Regarding the process of programming a CNC machine to cut production, answer the following.
a.) How is a resultant feed-rate calculated for a three-axis linear move?
b.) How is the resultant feed-rate calculated when a rotary axis is added?
c.) Express the maximum tool radius offset to execute the angled cut in the figure shown below. Assume R≥ X and X=0.0156 in.

4 *Electric Power*

4.1 Main Power-AC

An ordinary household light bulb burns sixty watts of electricity. Power for the bulb is delivered over a two-wire, single phase 115 Vac circuit. By comparison, an average size machine tool spindle motor is rated for 15,000 watts. For this kind of power transfer, electric companies need more efficient systems (three wires instead of two). For industrial machines in North America a balanced, three-phase 230 Vac service is common.[1]

After a good spot is found for a new machine, a licensed electrician comes in and connects the *main* electrical service. The wiring installation taps off the shop's main power bus-bars with approved, professionally installed fuses, conduit and wiring.[2] Branch wiring terminates inside the machine cabinet.

Main power for the machine is delivered over the three wires nicknamed R, S, and T. The correct gauge wiring is specified by the machine's maximum current rating. The diameter of the wiring needs to be large enough to deliver the machine's maximum current without losing the voltage amplitude. As a rule of thumb, the lug connections and wire diameters used on the main terminal strip should be equal or greater in size than the original factory wiring.

The main power installation needs correct *phasing*. Coolant pump motors, hydraulic pump motors and chip conveyer motors will reflect a phasing problem. Check the direction of rotation on all of the three-wire, squirrel cage AC

[1] The Small Motor, Gearmotor and Control Handbook, Fifth Edition, Bodine Electric Company, 1993. Local power levels vary between countries, overall machine specifications are designed to accommodate most markets. The certification standards also vary between countries, "UL" (Underwriters Laboratories Inc.) in the US and "CE" for products sold in the European Union.

[2] National Electrical Code® Electrical Systems Based on the 1996 NEC®, Michael I. Callanan, Bill Wusinich, American Technical Publishers, 1996.

motors. Many times when the phasing order is correct, the fan will turn in the direction of the arrow printed on the fan shroud.

After an electrician finishes hooking up the main power they will check the phasing. If any motors are turning backwards, the phases are changed. A standard three-phase, line-driven motor will rotate in the direction given by the phasing.

The motors driven with solid state inverters may have phase independent rectification, but may also contain phase dependent cooling fans. Electricians routinely check with the original equipment manufacturer to find out correct phasing connections and overall specification for the main power to the machine.

The electrician will also check the main voltage at the machine. If the line voltage exceeds the specified limits the electronics will fail. The electronics were designed to give a long life within certain limits—running near or outside these limits shortens their life.

In practice, the main voltage in a shop swings up and down. These power fluctuations come from the power company, or in response to other nearby equipment. During the night shift when many machines on the line are shut down, a high voltage may appear (like 250 Vac), even though the daytime levels stay within specs.

A step down transformer can be used to shift the line swing to within the safe voltage margin allowed in the machine specifications.

4.2   Step Down Transformer

Warranty contracts dictate that all new machine installations have the line voltage checked. When the voltage is too high, a step down transformer is added outside the machine. If the shop's power is known to be high, the cost of a transformer is included in the original quotation for a new machine.

A step down transformer is built of wires wound around a magnetic metal core. By adjusting the ratio of wire turns between the primary and secondary terminals, the voltage output becomes a scaled down replica of the input. Taps select the desired output. A transformer with a 240 Vac primary for example, has secondary taps for 208, 220, or 230 Vac.

The correct power rating for the transformer must exceed the maximum power demands of the machine. If transformers get hot, a licensed electrician should double check the installation. When locally available, the oversize, second-hand transformers do a great job at a reasonable cost.

Step-down transformers are mounted securely on solid stands that put some distance between the computer circuits of the machine and the powerful electromagnetic fields surrounding the transformer. No bolting the transformer directly on top of the computer cabinets. In practice, a step down transformer lowers the voltage and adds a measure of surge isolation and protection.

4.3   Power Conditioner

Opinions seem to vary on how "clean" the power to the machine needs to be. Companies selling power filters offer impressive before and after stories.

A power filter cleans the power by rejecting the unwanted frequencies in a signal. The expense of applying a filter to "clean up" the main power is very high. Smaller, low current units are less expensive and are used to filter an individual computer power supply or CRT power line.

People in the industry swear by the copper grounding rod driven six-feet into the floor next to every machine. Efforts at reducing electrical noise around a machine are welcome. The knotty subject of electrical noise is treated in the references.[3]

---

3 Surge and ground loop problems are reviewed by Dan Fritz, "Keeping Your Shop On-Line," Modern Machine Shop, March 1995, 104-109. See also the reference given in Section 7.9.

Decisions concerning power filters, ground rods, transformers, Star-Delta and Delta-Wye line configurations require the help of a licensed local electrician. The electric utility companies also provide professional line monitoring equipment, advice and analysis.

4.4   A Hundred Volts-AC

The main voltage service distributed throughout the machine is transformed into branches of secondary power—the first branch highlighted is called the *AC100* power.

A single-phase transformer mounted inside the control cabinet produces the 100 Vac power needed to power relays, solenoids, cooling fans and even the ordinary household wall sockets which are conveniently mounted on some machines.

The AC100 transformer is tapped to give a solid 100 Vac output for a range of service inputs. Commonly, the transformer tap wiring allows input options of 200, 220, or 240 Vac.

The machine schematics contain the secondary 100 Vac power circuits. Typically, these circuits take up many pages in the circuit diagrams. An electrician working with the factory can trace down troubles in the AC100 circuit.

4.5   Power Supply-DC

Every machine needs clean DC power to run the computer boards, feedback units and limit switches. A custom built metal chassis contains this DC power supply unit. The size of the unit is related to the current demands of the machine. Older machines have higher requirements; therefore, huge power supplies, with external cooling fins are required. The new machines use compact, efficient plug-in style units.

Computer boards require smooth DC voltages for their logic operations. If levels jump, the computer logic freezes up. Regulated DC outputs send the smooth, low *ripple* voltages vital for pulse counting computer and feedback circuits. Ripple is checked by a service engineer using diagnostic test equipment discussed in later sections.

The power supply converts a large input voltage (typically 200 Vac) into the small DC outputs (typically 5, 12, 15 and 24 Vdc). It also detects an alarm if something goes wrong. Whenever there is a power supply problem, the machine quickly shuts down.

Two independent alarm detection schemes perform the shutdown function. The *fuse lost* circuit is triggered when a fuse burns. The fuse alarm contacts are placed in series so if any one blows an alarm shuts down the entire machine. A visible flag, or status LED, is raised to indicate when a fuse blows. Older machines use replaceable fuses, while the newer machines use circuit status LEDs that automatically reset when the short is removed from the circuit. The machine will not run until the cause of a blown fuse is found and corrected.

If a circuit out on the machine shorts, the power supply blows an output fuse. Applying Ohm's law to the circuit shows a shorted load requires more current than the fuse will tolerate. The short is found and remedied before replacing the rated fuse. A power surge often burns the supply's input fuse. If this fuse keeps blowing, the entire supply is suspect.

Another type of alarm is detected by internal circuitry wired deep within the power supply. These alarms are the problems not detected by a blown fuse, but could still lead to computer damage or loss of control. If the *power lost* circuitry detects an unusual level drop, or voltage problems inside the unit, an alarm is sent to shut down the machine. The power lost alarms are usually very intermittent, and therefore very serious. A complete supply replacement is the prudent step.

The main supply for the machine is called the *internal* DC power supply. Any other "extra" power supplies are called *external* units. The external units may power up the machine initially or supply external circuits like machine limit switches. Further identification requires the specific elementary schematics of the builder involved and the builder's specific inter-pretation and assistance.

4.6   Axes Servo Power

The axis motor drivers receive two separate and distinct stages of electrical power. The first power received is for the computer logic circuits, and is often referred to as the *control-power* circuit. Next, the second power, or *main circuit*, is made available from which the heavy current for the motor is generated.

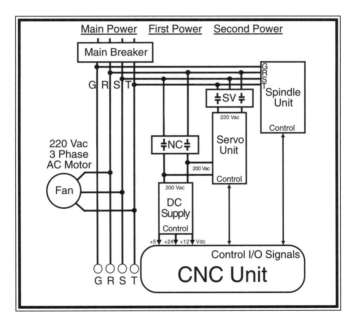

Figure 4.1 Power Distribution

In this way, the servo is checked internally by the computer logic before high level current is applied. This chain of events is critical. It doesn't make sense to send heavy currents into a unit with a problem. Likewise, when a service engineer is trying to isolate a problem they will not activate the main circuit if data is safely displayed at control-power only status.

The servo systems on most types of NC machines will follow a control-power then servo-power sequence.

Because servo problems are discussed so frequently around the maintenance network, good quick answers are usually "rattled off" by the OEMs over the phone. For specific servo power connections the factory supplied schematics are referred.

4.7   Spindle Power

From a power standpoint, the spindle is often a stand alone unit. When the spindle drive receives power, the lights come on, and drive status is made available. The spindle control voltages are generated internally by a small transformer and power supply. The spindle motor turns only after a set of interlock commands are given from the NC computer. This safety sequence is designed into the machine.

Spindles "charge up" and are best avoided because of the dangers presented, even after the power is off. The specific spindle power connections are in the factory supplied schematics and tend to vary considerably between the makers.

4.8   Power On Sequence

As already suggested, a sequence of events takes place when a CNC machine turns on. In the sequence, main power, first power and finally second power is established before a CNC machine is really considered up and running. This sequence of events is called the *power-on* sequence.

The sequence starts when the *main* breaker is flipped up, sending a distributed web of power around the machine cabinet. Next, the "on" button is pressed for the *first* time. After checking that things are in order, the DC power supply comes on, lighting up the computer displays and activating the control logic functions.

When the on button is pressed a *second* time, more things are checked; if okay, the main servo power is delivered. The elementary machine schematics show the circuits that perform this power-up function. With a lot of hard, hard study the proper sequence of events is identified from the schematics. It's a whole lot faster to use someone who already knows the sequence. They will do a few key checks, and from the results, know exactly what is wrong.

Old machines use *ice-cube* relays, hooked together to form a hard relay logic network. Old pinball machines also used hard relay logic. Troubleshooting these hard relay logic systems is always a challenge. The newer machines have put the power-up parts together on a compact circuit board predominantly controlled by NC logic.

The next three sections quickly explain common terminology used in conjunction with powering up a servo controlled machine tool. These terms help describe the stages in the power-up sequence. Later, the distinctions will become more important.

4.9   Main Power

When the main circuit breaker on the side of the machine is turned on, the machine is said to be at *main power* status.

The functions activated at main power status changes from machine to machine. A few cooling fans and status lights are usually shown in the machine schematics. After the machine has main power, the next step is control, or computer power.

4.10   Control Power (1st Power)

By pressing the green "on" button for the first time, a *check* is made. If the check passes, a small relay energizes for NC power. This in turn latches a larger magnetic contactor (NC-M), sending power to the machine's DC power supply. With the power supply energized, all the computer displays come up.

75

The "check" looks at the alarm detection from a host of sources, selected by the makers. If there is a problem, the control power is quickly switched back off, and everything blinks out. These events are intended to protect the computer, operator and machine from damage.

Assuming everything first comes up, the machine is said to be at *control-power* (or 1st power). Internally, every printed circuit board is running. Any logic problems with computer, servo, or spindle will show on the NC alarm pages. With no unusual alarms, clearance is given for the next level-up in the sequence called *servo power* (or 2nd power).

4.11   Servo Power (2nd Power)

With all the computer boards activated, the internal diagnostic systems are all running. Like getting a rocket cleared for takeoff, these systems decide when everything is ready for lift-off.

If all the tests pass, the second press of the "on" button will kick in the big servo power contactor (SV-M). If the heavy main current causes an alarm, the machine generally drops back down to first power, with a new alarm sitting in the NC alarm pages. When everything is normal, the machine reaches second power and operations like moving the machine around, zero return and automatic operation can then begin.

Whenever a machine will not power up, the OEM can determine at what point the sequence failed and offer a remedy. An engineer will never bypass or attempt to force the power-on sequence, damage will occur. When the machine won't power up there is a very important reason.

(Note: The power-on sequence for some machines happens all at once. With a press of the "on" button the machine blinks and clicks for a while and everything comes up automatically. These machines follow a power-on sequence, but the CNC-maker decided to automate its action.)

CNC QuizBox

4.1.1 What is the definition of power factor? What is the best power factor for a machine shop? What are the symptoms of a shop having a terrible power factor?

4.1.2 Sketch a one-cycle picture of the following voltages found on a machine tool.
   a.) AC100
   b.) "R" Phase
   c.) "S" Phase
   d.) "T" Phase
   e.) What is the instantaneous voltage sum of R, S and T?

4.5.1 Sketch a picture of three switch contacts in:
   a.) Series
   b.) Parallel
   c.) Which circuit can monitor three alarm contacts on one signal line?

4.5.2 Sketch a picture of three 1Ω resistors in:
   a.) Series, then find the equivalent resistance
   b.) Parallel, then find the equivalent resistance.

If you need help and further explanation for the exercises given in this book, go to "The CNC QuizBox" student area at the publishers website www.cncbookshelf.com

5 Servo Loop

5.1   Axes of Motion

On the machine, servos move things around, and the spindles provide powerful, high speed turning. Because the machine has so many axes, each is labeled with a letter from the alphabet. This conveniently separates and identifies the different servo systems.

Lathes and mills are two of the basic machine styles. A simple slant lathe uses two linear-axes named X and Z, along with a single spindle axis. By contrast, an expensive five-axis vertical machining center has three-linear, two-rotary and one-spindle axes. The abbreviated axes labeling is: X, Y, Z, *4th*, *5th* and *S1*, respectively. A full "5-axis" mill has programmable motion in a total of six simultaneous axes (including the spindle).

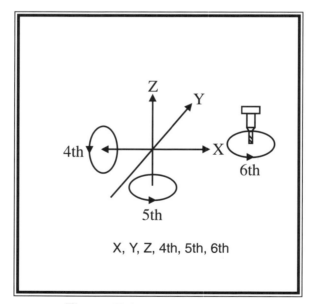

Figure 5.1 Motion in an Axis

Gordy picks up the phone and a customer says, "The Z-axis won't move this morning. The X is fine, but whenever you try to move Z, the machine shuts down. The machine gets the same alarms every time, and they won't clear, except by shutting off all the power. The Z servo-unit displays an alarm code meaning the Z-motor is being shorted, but it was checked by a motor shop, and they said it's good. This problem has happened in the past, but only intermittently—now it's happening all the time. Could the Z-axis ball screw be causing this trouble, or is it something else in the Z-loop?"

A one-axis system has less *freedom* of motion than the five-axis systems. The term *axis* only defines a direction of possible motion. The several axes and axis combinations applied to CNC machine tools can create complex permutations of motion. To help keep things straight, axes are classified by three subtle criteria: (1) the level of motion control; (2) the direction of the *pure*-axis; and (3) the resulting mixed axes movements.

To assign a level of motion control, consider a *full*-axis as one having complete position **and** speed control; a *half*-axis controls just the axis speed **or** position. For most CNC applications, a servo-axis requires full motion control. The spindle axis gets by on half-axis control.

A simple back-and-forth sliding motion is a *linear*-axis. A full-axis revolving 360 degrees is considered a *rotary*-axis. A spindle also revolves 360 degrees but is not considered a rotary axis because of its half-axis control. Combining the pure movement of two independent linear-axes allows motion in angled straight lines, looping circles, or any two dimensional path. Add another free direction of motion, for a total of three, and tracing helical coils or other three dimensional (3D) shapes is possible.

Synchronizing the movements of a spindle and a servo enables rigid tapping, constant surface speed cutting and return pass threading. The polygon cutting, passing live parts and lathe milling result from other combinations of synchronized, mixed-axis motion.

5.2   Overview

Modern CNC-mills run elaborate contour programs to build up the production molds for mouthwash bottles, egg cartons, aspirin tablets and so forth. Before entertaining the subject of multi-axis motion, a quick introduction to single axis motion is in order.

Consider how a machine simply moves 10 inches along the X-axis direction. Starting with a program command for X10.0 and block execute, the X-axis motor starts turning, accelerating up to a constant speed, while beginning to cover the requested distance. When ten inches has almost been traveled, the motor suddenly decelerates to an exact stop. The X-axis on the machine slid exactly ten inches from where it started, the motor speed, slide distance, holding power, everything—all scientifically controlled.

These incredible mechanical movements are performed by the axis closed-loop control systems. A full-axis servo contains two *closed-loops*, one loop controls the speed, while the other loop controls the positioning. Both loops are closed by the motion feedback. The speed or *velocity*-loop needs speed feedback; the *position*-loop finds its way using position feedback.

Both loops lock-in and hold their desired motion by quickly sensing and supplying minor command corrections. Corrections represent the difference between where the loop should be (command) and where it is (feedback). Introducing new motions into the loop is orchestrated by the position-counting and speed-commanding actions done inside the NC computer.

How both loops act in unison to provide the overall motion control solution is a challenging concept. Think about the simple ten-inch example, where speed and position was simultaneously controlled. What action caused both loops to finish moving and go to zero? Or does the machine really finish moving? The actions of a closed loop supply the answers.

The first modern CNC machines were mass marketed after affordable electronics could control all the necessary qualities of mechanical motion. These qualities, which are well defined in classical physics, are relatively unchanged in the college textbooks to this day. When marketable control electronics arrived, new answers appeared for the dusty old motion control problems.

Applying electronic control to mechanical motion is a subject for scholarly pursuit. The subject is taught at the universities to undergraduate and graduate students in Electrical Engineering. Prerequisites generally include the study of pure physics, mathematics and mechanics with special attention to control systems, digital circuits, computer architecture and software modeling of real-time systems.

The description of physical motion involves acceleration, velocity and displacement. Describing mechanical motion adds mass, force, friction, torque and inertia to the mix. Finally the electronic control of mechanical motion brings in the additional concepts of current, voltage, frequency, gain, stiffness and the dynamic characteristics like duty and ripple.

In the past, the OEMs would hire service people with high school diplomas and put them in the field for some "on the job training." Today, new service engineers have formal training from technical colleges or even university degrees in Electrical Engineering. While it's true that some motion control calculations depend on a formal engineering education, an effective compromise is reading the current articles and taking industry specific training from the many trade schools.

Familiarity with the distinctions between velocity and speed, position and distance, and linear and rotational, is helpful. Relating motor velocity to axis velocity, and axis position to motor position is handy. Also, whenever possible, concentrate the discussion of motion control to the four common quantities of power, acceleration, speed and position. Derive the other qualities of motion from this set, without becoming overly bogged down in the derivations.

In the last twenty years, the application engineers at CNC factories have steadily incorporated the newest motion control solutions. They are driven to beat out their competitors. Adding improvements over a twenty year period explains the roughly half-a-dozen clearly different versions, or generations of technology found scattered in the machine shops of today.

5.3   System Comparison-AC vs. DC

The systems for the electronic control and synthesis of mechanical motion have dramatically improved in the last twenty years. Some of the interesting technical milestones in this journey include the gradual transition from analog to digital control and the distinct change from the old *DC* (Direct Current) motors to the *AC* (Alternating Current) motors of today.

The DC technology first appeared with parts and circuits all of a like technology. Down the road, AC systems slowly started appearing, first in spindle applications and before long, a complete transition to the desirable AC motors. Together the AC and DC control styles open a wide umbrella over all the axes and spindle technology found in the last twenty years.

Those CNCs built twenty years ago could precisely control mechanical motion, maybe not as fast as the machines of today, but the underlying subject of motion control applies. Regardless of age, the objectives for a machine have remained relatively constant as the style and fashion of the electronics, motors and associated drives have changed. The nuts and bolts differences are only a minor hurdle once the key ideas applicable to both styles are considered.

To highlight the distinction between the DC and AC viewpoints, each chapter that deals with the servo and spindle loops is broken into separate sections. Within each of these chapters the features of the older DC technologies are first given in detail, then the newer style technologies, which still perform the same basic job, are given a brief update charting the improvements in overall performance. (The next section shows an oval symbol to remind whether the *newer* AC, or *older* DC systems are being discussed.)

All the machines running in the field, from the first-generation systems of the '70s, to the newest AC systems of today contain noticeable overlaps of control ideas and motion technology. This is a valuable insight for machine tool service.

5.4   Features of the DC Loop   ⬭ *Older Systems*

During the course of daily CNC maintenance, the subject of servo loops is an often told favorite. A brief introduction to loop action will help highlight the common ideas and terminology overheard in these discussions.

When a motor speed is selected by the NC computer, the servo amplifier receives this command and cranks out the necessary motor current. The motor springs to life, turning just long enough to complete the move, then stops. Meanwhile, the move dynamics were recorded by a precision motion feedback system. All the pieces within an active closed-loop must pull together.

The servo generates a powerful motor current, accurately adjusted to any speed. A transistorized DC servo-unit makes the motor current, always matching the speed requests with the motor's actual speed. Depending on the manufacturer, the unit making this motor current is called a *servo-amplifier, Servo-Pack, servo-driver*, or just *servo*, for short.

The motor faithfully reacts to current from the servo with rapid rotations and stable holding power. Because rotation is continuously monitored by position and speed sensors, should the motor ever step out of line, the stray feedback causes the NC and servo to quickly jerk the motor back in line.

5.4.1   D/A Signal

Older Systems

It's hard to imagine, but a speed command sent from the NC computer can actually move and hold the motor shaft in the exact, right *position*. To obtain this accuracy, the command is generated by a digital-to-analog (D/A) conversion chip mounted on the computer boards. The computer understands digital and the servo understands analog, so the computer commands the servo by converting strings of digital data into highly controlled analog speed commands that the servo recognizes. The speed command receives the nickname *D/A command*, (short for computer-generated, digital-to-analog speed command).

Assume a ten-volt D/A command runs the motor at its maximum rated speed. By reducing to five-volts, the motor slows to half its rated speed. A positive DC voltage commands clockwise rotation, and a negative polarity reverses motor rotation. A smooth and steady command causes a constant motor rpm. The speed of the motor is always in proportion to the DC level of the D/A command.

Returning to the previous discussion of a ten-inch move: what was the shape of the D/A speed command? First, the acceleration of the motor (a rise in the D/A) is followed by a steady rpm level, ending with a falling deceleration back to zero. So, from zero it went up, flattened out and then came back down to zero. A *signature* D/A command shape exists for this and every closed-loop motion.

Shape signature is determined by two things: the type of move commanded and the resulting position feedback encountered by the motor's response. The computer is always counting feedback, crunching out the answer for what becomes the final D/A shape. Signature command shapes for rapid and feed style moves are discussed after a few more D/A preliminaries.

The smallest D/A command, or the D/A minimum increment, is issued when the axis is out of position by exactly one "friendly" pulse. Stated another way, the smallest D/A equals the smallest position move of the machine. To complete the smallest move, the overall loop position gain (K_p) must be high enough for the machine to respond to this smallest of D/A speed commands.

For every unsatisfied or *undistributed* position pulse, the D/A command climbs another small step in the total command staircase. As more undistributed pulses back up waiting for a rotation, the D/A command climbs ever higher, causing the servo to send a bigger and bigger motor current. When the axis finally starts moving, the completed feedback pulses come in and lower the staircase step-by-step until every pulse has been distributed. In this way each command pulse is matched by a feedback pulse, and the move is completed when the D/A returns to zero.

In practice, the D/A hovers around zero by a few pulses. If it pushes too far away from zero, a simple adjustment removes the command drift (not uncommon in aging analog D/A circuits).

5.4.2   Velocity Loop

Older Systems

The servo makes current for the motor in proportion to the D/A command level. For longer moves, like the ten-inch example, the motor actually reaches and runs at constant speed for a time. To keep the motor speed smooth and constant, a closed velocity loop is set up. The servo compares the D/A command (the speed desired by the NC computer) with the actual speed of the motor. (The actual motor speed information is faithfully generated by a speed sensor tied to the spinning motor shaft.)

Keep in mind, this spinning motor is hooked up to some very heavy cutting machinery. The load on the motor changes. If the load rises the motor will naturally want to slow down. Now that's a problem—the NC computer wants constant speed, but the motor wants to slow down. A *lag* develops between the actual and commanded motor speed. Like driving up a hill with the cruise control on, the servo loop notices this lag and responds by bumping up the current to get the motor up that hill.

As long as things are kept within specifications, the correction between commanded speed and actual speed is performed quickly by the velocity servo-loop. The rate of speed correction is known as the velocity gain (K_v). Adjusting the gain too high leads to unstable oscillation, while adjusting it too low results in sluggish operation.

Position lag or error increases as the motor is loading up. If the load becomes too heavy, of course, the motor will stall (stalling the motor is bad for the machine). Servo lag turns out to be a powerful diagnostic for machine tools. The error pulse display and subsequent calculations are used to evaluate the health of a servo loop. Further discussions of lag and error are included later in Parts Two and Three.

5.4.3   Position Loop

A previous discussion of positioning loops was simplified with the help of "friendly" position pulses. Now a few more details and repetitive examples are added to better develop the idea of position loops.

The motion control of an axis is governed by three important quantities: the motor's rated speed, the rated power and the minimum increment move. Physical motor specifications govern the rated speed and power, and the minimum move or increment is determined by the position loop specifications.

The minimum increment is given in the basic machine specifications. A typical value for machine tools is a metric value of 1 micron (0.001 mm). Machine shops in the U.S. predominately use the inch system in which the smallest move is a *tenth* (one ten-thousandths of an inch, or 0.0001 in).

If a single command pulse (minimum increment = 0.0001 inch) results in the smallest possible move, a one-inch move requires exactly 10,000 of these pulses. Now remember, these distances are linear movements caused by a motor's angular, twisting rotation of a ball screw. The relationship between a full revolution of the motor and the resulting linear move is simply the pitch of the ball screw, plus any gearing involved.

The computer keeps track of the positioning by counting positioning pulses. Suppose the operator executes the smallest move of one-tenth (not 0.1, but 0.0001 of an inch) from the operator's panel. In this scenario, the control sends out a single pulse command to the servo for a small motor rotation. When the motor rotates, a single equivalent pulse of feedback is sent back to the control. The move is finished. The computer has sent out one command pulse, received back one feedback pulse, and the pulse sum is back to flashing zeroes. (Note: It's easy to confuse feedback resolution with the friendly pulse. The minimum resolution distance of the feedback unit is smaller than a friendly loop pulse by a design scaling factor.)

Now return to the ten-inch move. First, the 100,000 (10÷0.0001) pulses are commanded. Then the motor rotates until the accumulated feedback pulses total 100,000 at which time the positioning is complete. When command pulses exactly equal feedback pulses, the positioning is complete, and the distribution of pulses is finished. For each one out, one came back—this is the essence of a closed positioning loop.

5.4.4   Two Loop Description

Every time the NC machine moves an axis, two different loops are simultaneously "closed." These are the loop of position and the loop of speed. This situation is common for motion control applications. Closed position loops provide accurate positioning down to the minimum increment. The closed speed loops provide for smooth running of the motor.

Careful examination of the elementary schematics reveals the wiring for these two loops. The speed loop consists of the servo-amplifier, axis-motor and speed feedback. The position loop is closed by the NC computer, servo-amplifier, axis-motor and position feedback. The two loops share a common signal—the venerable D/A speed command.

The speed command is key to the entire puzzle. Return again to the ten-inch move example. The machine moves from its current position to a new position ten inches away. During this move, the motor speed ramps up to a fixed speed limit while the 100,000 position pulses are being distributed.

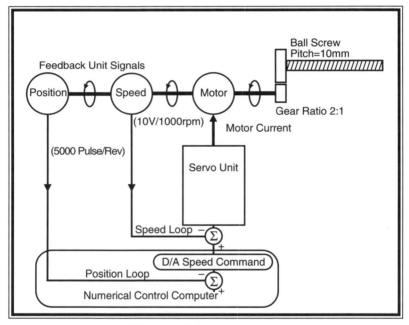

Figure 5.2 Two Loops

The NC computer sent the speed commands to accomplish this result, however, the speed and position loops made the motion feedback to close the loop. In Part Two, loop detail and operation is checked using the Diagnostic Tools.

5.4.5   Feed and Rapid Modes

When the machine is casually moving around, perhaps moving away to a safe position for tool change, but not actually cutting metal, a fast mode of travel is used. This fast positioning mode is called *rapid* mode. In 100% rapid, the machine runs at its maximum designated speed limit. The zip, zip, zip rapid speeds on the newest machines approach **forty** meters per minute (or 1575 in/min). That is some fast moving metal, and with the introduction of linear motors they keep getting faster. The maximum rapid speed is one of the basic specifications salesman use when bench-marking a machine.

The highly controlled tool speed necessary during metal cutting is named the cutting *feedrate*. The correct feedrates, or feeds, are directly typed into a part program using *F-codes*.

While cutting through stainless steel is an inherently slower process than ripping through soft aluminum, the exotic materials like titanium and Iconel require ever vigilant attention to the specific cutting speeds and tooling.[1] The correct cutting rates, or feeds, are well documented in the machinist handbooks and tool insert buying guides.

The feedrate, or unit of speed, is given in either inch-per-minute or inch-per-revolution formats. The maximum inch-per-minute *time-feeds* on a standard machining center are around five meters per minute (or 200 in/min). This, combined with the maximum spindle rpm, is ample for most machining methods. For faster, high-speed machining the speeds and feeds are bumped up with expensive, optional OEM features. (Reviewed in section 9.2.4.)

The inch-per-revolution *rpm-feeds* are a little more difficult to summarize as they describe a relationship between **two**

---

1   Advanced Machining Technology Handbook, James Brown, McGraw Hill Text, 1998.

moving bodies. The rpm-feed relationship occurs at the point of contact between a spinning metal part blank and the moving cutting tool. The idea of an rpm-feed improves with a brief discussion of machine lathes.

Cutting parts on a turning center involves using the "lathe jargon" for the geometries involved. To begin, a lathe spins a part in the spindle while the tool makes a brisk approach to this spinning part and takes out a cut. The jargon includes the part-*face*, part-depth (*Z-depth*), outside diameter surface (OD), inside diameter surface (ID) and the *center-line* of rotation. In lathe-speak, for a part spinning in the chuck, the OD moves faster than the ID because it's farther from the centerline. Likewise, the surface speed increases when moving up the face, from the centerline to the OD. The lathe programmers see the whole idea as simple conventions in machining.

A lathe's cutting tool has tiny replaceable inserts. These tool inserts are designed to cut through specific types of metal within a fixed range of recommended feeds. A constant surface feed gives the optimal result. Following the machinist handbooks or tool company's recommendations for *feeds and speeds* gives longer insert life, good cutting finish and better chip removal.

With that brief introduction to lathes finished, we return to a discussion of the inch-per-revolution feedrates. The rpm-feeds are specified in the program using *F* (Feed) and *S* (Spindle Speed) *Codes*. A constant surface feed naturally occurs along the Z-depth at constant spindle rpms. However, cutting along the face of the part requires a special mode called constant surface feed (G96).

In this mode, the cutting tool tip and part blank are kept in a constant feed relationship by rapid computer adjustments to both the spindle and servo-axis speeds. The computer looks at the numerical distance, from center line to actual position, and quickly computes the right speeds for both the spindle and servo axes. Watching the machine run in G96 mode shows the spindle quickly picking up speed as the tool inches down toward the center-line. At the center-line, the spindle is running at maximum speed (unless a speed clamp, G50, is programmed).

With this brief understanding of feeds and rapids, the key subject of D/A command shapes is revisited. The rapid mode uses quick, straight line ramps during the *accel/decel* portion of the D/A speed command. Feed mode uses a gentle exponential curve in the accel/decel regime. Pictures of these signature command shapes are sprinkled around in some of the CNC operator and connection manuals.

The D/A speed command is created by a *motion generator* circuit inside the CNC computer. The motion generator sends out the correct D/A shapes for a rapid or feed mode. In rapid-mode (in/min mode), the motion generator uses a regular time clock. In feed-mode (in/rev), the motion clock is supplied from a spindle rpm counter or *encoder*. Calling for a feed-per-revolution, with the spindle stopped causes the machine to sit and wait forever. Once the spindle starts turning, the axis will start moving at the *F-code* programmed in/rev rate.

To review: A rapid-move generally is used for quick positioning of the machine; a feed-move is used when the tool is cutting metal, and fixed, controlled surface-feeds are required. Both cases rely on position feedback for distribution.

5.4.6   Loop Specifications

Every servo loop has a list of performance questions. How fast, far, stiff and accurate? How much power? When the axis servo system was originally designed, the choices for mechanicals and electronics combined to answer these overall loop questions.

Several issues contribute to the final loop specifications. When the weight of the machine mechanicals are shot back-and-forth, side-to-side and up-and-down, each axis on the machine has a value of inertia to overcome. The axis motors can only supply so much rated power and speed. The stiffness or rigidity of the different machine castings also dictate machine dynamics.

From a service standpoint, when a machine is running at *original factory specifications*, it obeys a set of verifiable guidelines. Checking these guidelines is done using the machine diagnostics. The display of the accumulated loop lag or *error pulse*, at a given motor speed provides valuable clues to the overall happiness and performance of the servo loop. External diagnostic equipment, like oscilloscopes and data recorders can capture the signature shapes of the speed command and motor torque, giving another effective analysis of the overall loop. In Part Two, the internal and external Diagnostic Tools are introduced to help verify loop specifications.

5.4.7   Alarm Detection

Older Systems

In the event of a problem, the servo-loop has its own built-in alarm circuits to protect itself and the entire machine. Each of these alarm detection circuits are vital to overall CNC safety. When alarm conditions are captured by the servo, the heavy motor current is instantly blocked off, and the NC computer is informed of a problem. The NC reacts by shutting down the entire machine, sending instructions to the other servos to quickly block off their motor currents as well.

Alarms are detected when: (1) fuses blow; (2) motor current is excessive; or (3) internal servo signals are faulty. The makers have worked long and hard designing alarm detection schemes that will quickly stop a machine if any of the servo-loops begin acting up.

One method to instantly cut-off the motor current in response to an alarm is accomplished by simply turning off all the servo-amplifier drive transistors at the same time. This condition called *base-block* effectively stops the motor current by reverse-biasing the base leg of each main motor current transistor. (Base-block performs a valuable safety feature for all types of servo-driven electric motors.) After the cause of an alarm is remedied, the base-block is released. The unit is then powered up, and normal operations can resume.

Whenever an alarm is present, the reset push-button on the servo-rack becomes dangerous. Pressing the button with the power on invites the possibility for sparks, smoke and flying pieces of shorted components. Remember, the servo originally captured the alarm and shut itself down—by pressing on the reset button, the alarm is temporarily cleared and the main base blocked current returns. If there's a short in the servo-unit or motor, an inrush of current takes place with fiery results.

Shutting down the entire machine and powering up normally, in stages, is safer. The point at which the alarm happens is determined. Is it at main? First? Second? On axis command? Every time a faulty servo is powered up (or reset!) a new risk for additional damage is presented. This situation doesn't allow for messing around, data is taken one time, the call for service or replacements is made. Blown servos are replaced around the business almost like fuses.

The oldest servo systems don't have visible alarm lights; only closed, internal alarm contacts tell the NC computer about a servo problem.

Newer DC units show status by lighting up little LEDs mounted to the front servo PC boards. Next to the base of each alarm light are a few abbreviated letters indicating the alarm condition: OC for over-current, OL for over-load, FU for fuse, etc. Sometimes the useless descriptions like "3LED" must be taken to the factory schematics for interpretation. The possible causes are found in either the servo-unit elementary diagrams or the alarm documentation for the machine. On these older units the OEMs have the good answers given out a hundred times before.

5.4.8   Input and Output Status

Servo-units and the NC computer always maintain two-way communication. The communications monitor alarm and status on both ends of the link. Recall the signals sent moving about, in and out and between units are called I/O-Signals. Using I/O-Signals, bi-directional status and alarm conditions are conveyed around the loop.

Status monitors communicate the normal status of everyday events: motor is turning, motor is stopped, and so on. Their change is normal whenever the servo-loop encounters a change in its running status.

If the servo falls into an alarm condition small contacts close telling the NC computer of the event. A servo alarm signal is an *output* from the servo-unit and *input* to the NC computer.

As mentioned, the status and alarm signals are two-way. A faulty computer will stop the servo and vice versa. Finding problems within a servo-loop requires checking the status of the servo I/O-Signals. Looking at this I/O data requires the internal and external Diagnostics reviewed in Part Two.

5.4.9   Run Enable

For safety and control, motion anywhere on the machine must always satisfy the motion safety interlocks. The terminology varies a bit, but the salient idea is captured by the concept of "run enable."

The run enable condition for older machines checks the D/A command, *forward* command and external *run* command to determine if motion is allowed. If any of these signals are missing, the axis isn't going to move. If the motor is turning when one of these conditions is lost, it comes to a halt.

The run enable interlock keeps the machine safe. Some OEMs will beef up their interlocks. The possibilities include: removing the servo power, base-blocking the servo, issuing sequence alarms, or triggering NC data that stops the faulty motions. The OEMs must help locate the myriad of conditions leading up to a motion interlock.

5.4.10   Servo Type

The cabinets mounted behind the machine keep all the electrical components in a relatively clean and temperature controlled environment. Inside these sealed cabinets are mounted the servo-racks. To help identify the correct servo for the correct motor, inspect the screw terminals for labels and lettering (i.e., *SVX, SVZ, AX, BX, AY, BY,* etc.). It's safe to assume that every controlled-axis motor on the machine must receive current from a servo-unit.

The servo model type is printed on the amplifier chassis or boards—somewhere. The complete type and serial number of the servo is written down before disturbing any connections or adjustments. Complete nomenclature is vital for ordering the correct replacement units, or if special procedures and documentation are required. Servo-units of different types are generally not compatible.

The numbers in the description usually reflect the unit's kilowatt power rating. If a unit is clearly burned, the model type is called in to facilitate a replacement. The availability of replacement units can take a few hours, or *forever*, depending on the unit.

Finding the same unit on a different axis lends a possible option to the repair (see "swapping" in section 12.2.7). As a safety precaution all servo compatibility and replacement information is received beforehand from the manufacturer. A good fax from the OEM is used to document a servo loop repair.

5.4.11   Replacement

Replacing a blown servo is possible using a Phillips screwdriver, but a few things are considered before rushing into a replacement.

To avoid mis-wiring, a careful drawing of the original system is made before anything is touched. The jumper set-

97

tings on the drive and the number of wires leaving every terminal is included. The settings of any variable potentiometers mounted on the circuit boards are also important. It's made doubly sure, before disturbing the original system, that a return to original is possible.

A very common characteristic of closed loop systems is how every component in the loop is related and is possibly suspect. When problems start passing around the loop, a fix becomes more difficult. A bad motor, bad feedback unit, bad incoming power source, or machine crash can by themselves short out a servo unit. Other loop possibilities include an overheated electrical cabinet, excessive mechanical load on the motor, a shorted motor wire, or even a defective DC power supply outside the drive.

If replacing the servo fixes the problem, great! The first step was the last step. Otherwise the investigation will expand to every piece in the loop until the culprit is found and corrected. To avoid burning up expensive new components, everything in the loop is individually checked and qualified.

5.5   Features of the AC Loop   ⬭ *Newer Systems*

Putting AC motors on CNC-tools was hailed as a tremendous improvement. The early changeover from direct current to alternating current systems was gradual. First the new AC spindle systems were matched with the old DC servos. Finally the transition to complete application of the new AC motion control technology occurred throughout the machine.

The definition and description of AC servo operations follows in the same progression as the sections just given for DC servo operation, so repeating the same ideas is hopefully avoided.

By no means is a complete description of every new AC system attempted, or even possible, only some of the interesting new highlights and developments are reviewed here. In some cases, the early AC technology uses the same loop schemes applied to previous DC systems. Factories establish where in the playing field a newer machine is playing. They deliver the specific procedures and service answers to keep the service quality high.

5.5.1   D/A Signal

In the last few years the standards for "encoded" speed commands have expanded.[2] The analog concept of D/A remains valid, but may no longer be directly measured. Special test monitor boards obtained from the factory must decode the encoded servo loop signals. This special test equipment must supply the D/A, motor speed and torque signals in a check-pin format suitable for external test equipment.

The signature shapes of the D/A-speed command now have more choices like: step linear, S-shaped and step exponential. The cross-over points and shape coefficients are more robust with more devoted parameter and setting addresses.

(Review Section 5.4.1; check for applicable systems using the OEM's help and the original schematics.)

5.5.2   Velocity Loop

Motor speeds on the AC systems are available to higher rpms. Expanded filter and gain controls reduce shock and resonance in the speed loops. The speed feedback data is decoded from motor pole-sensors, pulse count frequencies and resolver signal frequencies. Gone are the days of the trusty old tacho-generator.

Some control makers moved the function of speed loop off the servo-units and into the motion modules of the NC com-

---

2   An interesting review of further efforts at standardizing CNC data transfer schemes is given by Golden E. Herrin, CIM Perspectives "Progress of SERCOS, " Modern Machine Shop, May 1996, 146-148.

puter. These servos are a good example of module slave amplifiers.

Still others installed smart servo-units in limited applications. These servo-units are programmable to handle stand-alone functions, offering a complete speed and position loop for the motor. A burst of serial RS232 data or other standard I/O directs how they are utilized. Very complicated to service, they require detailed application manuals from the OEM to understand their all-in-one operation.

(Review Section 5.4.2; check for applicable systems using the OEM's help and the original schematics.)

5.5.3   Position Loop

Digital control of the new position loops has eliminated drift. A fact reflected in the position error pulse display, now the error is steady while sitting, and keeps in tight formation while moving one or many axes. The use of software and digital tuning has changed the behavior of error pulse on the machine. The drift components introduced by analog D/A converters is now directly observed and removed by digital display-removal procedures.

Using a new generation of feedback units, the positioning systems are separated into *absolute* and *incremental* positioning schemes. The absolute systems remember the position of an axis after the power is shut off. These smart encoders "remember" by sending in serial position data at power up; after initialization, they act similar to the older incremental units.

More optional features are applied to new positioning loops, such as higher speed positioning, feed forward control loops, data form compensations and circle cutting adjustments.

The direct transfer of positioning segments using pre-processed program files speed up the NC motion processing time, allowing higher cutting speeds. The expanded maintenance pages capture trends of accumulated error pulse for multi-axis interpolation and synchronization adjustments.

(Review Section 5.4.3; check for applicable systems using the OEM's help and the original schematics.)

5.5.4   Two Loop Description

Most CNCs today perform the majority of loop processing inside the CPU rack. The servo-units are economical slave amplifiers. Improvements to servo-loop systems include both software and hardware refinements. These changes allow more detailed motion parameters and settings to accommodate a wider range of motion control solutions.

Some of the newer loops need special factory test equipment to check loop signals like speed, torque and feedback. Recently factories have begun to move away from this specialized field test equipment to ease their own service demands.

(Review Section 5.4.4; check for applicable systems using the OEM's help and the original schematics.)

5.5.5   Feed and Rapid Modes

Improvements include extended feeds and feed override settings, increased rapid rates and extended linking of multiple-axes interpolations and synchronizations.

(Review Section 5.4.5; check for applicable systems using the OEM's help and the original schematics.)

5.5.6   Loop Specifications

Older, mixed (analog/digital) systems have several jumpers and loop adjustments potentiometers that need to be checked during and after replacement of loop components.

Newer, completely digital systems hold their loop specifications after replacement if the original software and setting parameters are maintained. There may be found one or two lonely address jumpers.

Dedicated maintenance pages on the newest controls can display in real time a servo loop's cutting load percentage or motor-phase balance. Lists of the user-level servo adjustments are increased and understanding their use vastly more complicated. They should not be changed without factory help.

(Review Section 5.4.6; check for applicable systems using the OEM's help and the original schematics.)

5.5.7   Alarm Detection   *Newer Systems*

Early AC axis systems have digital alarm displays in a seven-segment alarm code format. The alarm codes are deciphered using the alarm-code lists supplied with the machine. Many types of alarms are detected, such as the loss of a power phase, regenerative braking troubles or a tripped off servo breaker.

Newer machines have since moved the alarm displays off the individual servo-units and into the main CPU rack. Detailed alarm pages display servo loop problems on the main operation panel. These servo units still detect certain alarms, only now they have no need for the redundant individual alarm displays.

(Review Section 5.4.7; check for applicable systems using the OEM's help and the original alarm code lists and schematics.)

5.5.8   Input and Output Status   *Newer Systems*

Expanded volumes of information are exchanged between the CPU and servo loop. As an example, the CPU monitors the load inside the servo loop. This information can be displayed on a bar chart for each axis, or even used to model mechanical systems in software. The electronic clutch is a good example—if the loop load climbs too high, a software clutch trips out, stopping the machine. Resetting or re-calibrating a clutch in software is much easier than climbing behind the machine to spring together slipped mechanical clutch plates.

Four other notable improvements to the I/O control of the servo loops include:

1) Options for connecting a machine's internal I/O data with an external PC computer for cellular networking applications.

2) I/O signals are detected using optically isolated, automatic, level-detection circuits.

3) Quick jumper settings and rotary switches simplify I/O address selections.

4) New schemes for I/O communications support an increased number of signal addresses. Use of optical and serial communication protocols and shared data-bus formats are provided to simplify wiring.

(Review Section 5.4.8; check for applicable systems using the OEM's help and the servo application manuals and schematics.)

5.5.9   Run Enable

The concept of using a run enable signal to ensure safe motion control is applied to AC systems.

(Review Section 5.4.9; check for applicable systems using the OEM's help and the original schematics.)

5.5.10 Servo Type

Many servo-unit parts may look the same, but the versions are not compatible. When calling the OEM for assistance, refer carefully to both model type and system version.

(Review Section 5.4.10; check for applicable systems using the OEM's help and servo application manuals.)

5.5.11 Replacement

Replacing parts in the mixed (analog/digital) AC servo systems requires OEM advised jumper and potentiometer adjustments. Factory supplied exchange procedures, written for the exact version under consideration, are as important as ever.

New servo units use fewer screw terminals, and the chassis size is smaller and more manageable than on earlier units. The new replacement units are normally set-up to look like the original. Paint-locked factory potentiometers are not touched.

The servo loops using all-digital systems use software parameter settings. These "smart" stand-alone units may have complicated initialization procedures best attempted by the responsible machine builder. The latest procedures fresh from the factory are best performed under new machine warranty.

(Review Section 5.4.11; check for applicable systems using the OEM's help and the original schematics.)

5.6   Servo Motors-DC   *Older Systems*

The design of old axis motors are varied, one style is the *cup-motor* design. In this design, a motor shaft connects to a molded Epoxy cup containing all the armature wiring. Surrounding this cup armature are powerful, permanent field magnets. These motors are both durable and long-lasting, especially if the brushes are periodically replaced, and the motor cavity is kept clean.

A close tolerance flange builds the linkage between the motor case and machine. The motor shaft is mated to the ball screw using a mechanical pulley or geared arrangement. The mechanical couplings fitted by the machine builder are found in the machine drawings. Some installations are tricky to figure out.

The motor case must not be disassembled. If the motor armature is removed the permanent field magnets are disturbed, rendering the motor useless. An opened motor never has the same power as before, typically, a bouncing motion results if the motor is put back on the machine. Re-magnetization of disturbed motor magnets, to factory original levels, is not possible in the field.

A young assistant of Gordy's, whose name is Kenny, gave a customer a stern warning against opening an old DC motor. Kenny, a fresh and eager serviceman, took pride explaining his story. The customer, on the receiving end of this story, was a kind southern gentleman who quietly listened, as southern gentlemen are inclined to do. After a brief silence, the customer began—"Well, I've been fixin' these machines a while," which was true— he was in charge of about thirty mostly new machines running around the clock—"And, I have to agree with you; the key is disturbin' the fixed magnets." Another silence. Now with a wink to Gordy and a small grin, he says, "What we do down here is cut a steel plug the size of the armature and slide it in while the armature comes out. That way the magnets don't get disturbed in the first place." Kenny remained silent, he had quickly switched from being teacher to student.

The motors receive DC power from two heavy gauge wires routed into an electric junction box mounted on the motor case. Power is directed into the spinning armature by long carbon brushes held stable in metal track brush holders. Commutator wear or discoloration is observed through a small removable inspection cover. Fingers, chips, dirt and oil are kept away from the motor if the inspection cover is removed, any contamination will ruin the motor.

5.6.1   Specifications

A metal nameplate riveted to the side of a motor contains the printed motor type and running specifications. Setting an old machine back to original factory specification relies on the data written on these greased-over nameplates.

Plates typically contain the motors rated running speed, power and current. The current ratings for thirty-minute and continuous running are handy for service investigations. The power ratings for a motor are more popular among machine salesmen and owners for making comparisons to other competing pieces of machinery.

5.6.2   Feedback

Motion feedback systems on vintage CNC machines are either linear-mounted *induction scales* or motor-mounted *ro-*

tary encoders. Since scales are less common, only the motor-mounted rotary units are considered. The important thing to remember is that the two motion quantities, speed and position, must somehow be derived regardless of the style or signal format of the feedback unit involved.

Encoder signal formats are much too specialized. As an introduction, one simple, *direct signal* feedback unit is discussed. All the other resolvers, scales and assorted encoder/decoder signal schemes are left to their OEM's application.

When an axis motor rotates, accurate information for motor position and speed is captured by a miniature, shaft mounted feedback unit. Within the "direct signal" feedback unit are two separate sections to detect and send the speed and position information.

Inside the first section is found an etched, optical rotary scale monitored by small infrared diodes. As the motor moves, the scale's finely-etched lines pass by the infrared detectors, generating a counting pulse that accurately reflects the motor position. A small printed circuit board captures and amplifies these positioning pulse signals. Contamination like oil or dust found on the optical scale will disturb the positioning count of the machine.

One full revolution of the motor gives an exact pulse count specified by the feedback unit model type. Typical resolution values include 2500 or 1500 pulses per revolution. The actual count is determined by the number of lines and channels etched on the glass scale, plus any electronic pulse division techniques.

The resolution of the feedback unit is compensated, or scaled, by the computer. Factories can change around NC parameters and servo settings to accommodate different feedback resolutions. The "friendly" pulse depends on this scaling being properly calculated.

In an adjacent compartment of the feedback unit is found a miniaturized *tacho-generator* (TG). Like a motor in reverse, this tiny DC generator has two miniscule brushes riding on a wire-wound commutator. As the motor runs faster, the generator puts out a higher voltage. This accurate DC voltage (called the TG signal) is directly proportional to the rotational speed of the motor. As the brushes and commutator become worn, the speed control of the motor becomes erratic and jumpy because the *TG* signal itself begins to jump.

A flexible access plug, found on the mounting body, gives access to the feedback coupling screw, making removal and replacement relatively straightforward. Feedback units are replaced in the field after a good replacement procedure is faxed from the factory. A factory installed feedback unit has a thin silicone rubber layer spread between the motor and feedback unit to better seal out the oil and dirt.

5.6.3   Motor Maintenance-DC

Motor life extending over ten years without special attention is common. Of course, simple preventative maintenance will help extend the motor's life.[3]

A neglected motor will short out from the excessive build-up of carbon brush dust inside the motor, this dust allows a current path to ground. Unfortunately, in the gallant effort to remove this dust, it may only be moved to another spot which causes a complete motor burn up at the next power on. As a precaution, the replacement motor stock is always checked before working on a neglected motor. If none is readily available, the motor cleanings are held off until stock levels rebound.

3 Maintenance Engineering Handbook-5th Edition, Section 7.21- Brushes and Commutators (D-C Motors), Lindley R. Higgins, McGraw Hill, 1995.

The cleaning begins by wiping away all the metal chips and loose oil in the area surrounding the motor, one stray chip can ruin the day. With the main power locked off, one motor is worked on at a time. The brushes and inspection covers are removed to a clean, safe place. The carbon dust is blown out using clean, dry, compressed air. Solvents are avoided for cleaning a motor unless the OEM specifically recommends differently. Breathing in the carbon dust cloud that comes from blowing out the motor cavity is to be carefully avoided.

Care is given to the old brushes so that oil is not absorbed from dirty fingers. Also the face of the brush is inspected for evidence of rocking in the brush holder. (A shorter brush moves back and forth in the holder every time the direction of the motor changes.) If the brushes skip while the motor is running, a spark or arc occurs, which can kill the motor.

Worn brushes are best replaced with new ones from the original factory source. The new brushes are ordered using reference to the motor model type, old brush dimensions, brush style and brush material. Once new brushes are installed and everything is back together, the axis motor is run very, very slowly in handle mode. Initial servo alarms reflect excessive current. When this happens everything is shut down, and the entire brush installation is re-checked.

After several slow motor revolutions in handle mode, the motor is run at a slow feed using a small test program. Every fifteen minutes the motor is speeded up in gradual stages until full speed works and the brushes are comfortably reseated. After everything is finished and completely checked out, the service moves on to another motor.

With periodic cleaning and fresh brushes, the life of the motor is extended, but as pointed out, there are risks. Before checking or cleaning an antique motor be sure the replacements are made available, just in case it should mysteriously short out. Cleaning motors is a big responsibility; if something goes wrong, an old machine will stay down for months waiting for a rare replacement.

5.7   Servo Motor-AC

The best thing about AC motors is that there are no brushes. Found inside AC motors is a spinning part (rotor) and a stationary part (stator) with only one thing to wear out: the motor bearings. The motor size packs a powerful punch compared to the older DC motor styles.

Burning out a quality motor is uncommon, unless the spinning part physically bumps into the stationary part. More often, a leakage of some coolant or oil into the motor cavity causes failures. Occasionally, a motor shop will successfully bake out a wet motor.

The motor type is printed on a tag stuck somewhere on the motor case. The motor description includes the flange type, shaft taper, oil seal, key-ways and power-off brake units. The power, speed and special motor traits are also coded into the factory's nameplate description.

Connected to the back of the motor is a sealed encoder. When this encoder goes on the fritz, the entire motor is often replaced. Encoder replacement in the field is touchy, and doesn't necessarily guarantee the same life as in a completely new motor-encoder combination. The service designs vary, so check with the manufacturer for the least expensive options.

5.7.1   Specifications

Modern AC motors have their complete specifications written in beautiful, widely-distributed factory application manuals. Complete loop adjustment and application is written up in the manuals. Everything is included, from power ratings to the physical dimensions of the mounting hardware. Factories provide these manuals to increase their sales in other motion control applications. A copy is generally supplied with the machine's original documentation, or is available by calling on the factory sales representatives. Both motor and servo are usually documented and sold as a matched set.

5.7.2 Feedback

There are several new schemes for tracking the speed and position of a motor shaft. Advanced design and manufacture of both optical and magnetic systems are now in the field. Careful alignment and testing is done at the factory before sealing the units. Field reproduction of these methods is difficult, if not impossible.

To review, the new positioning systems fall under two classifications: incremental or absolute. Incremental positioning includes the systems of old. Every time the machine is turned off, a new position initialization is required. This initialization is simply the zero return, first thing in the morning at power up. Absolute systems automatically perform this function by sending a serial packet of position information to the computer at power up. After initialization the absolute encoders send incremental position updates (like the older units) in response to machine movement.

Modern absolute encoders have resolutions that reflect binary data, like 8192 pulses per revolution (or 4096, 2048, or 1024). The simpler incremental units have resolutions like 5000 or 2500 pulses per revolution.

Motor feedback information is also sent using many new, hard-to-imagine schemes of detection (things like resolvers, pole and phase sensors and pulse frequency shifting/counting). All different methods are developed to reduce manufacturing cost and increase the reliability. There's no telling what will show up next. (Review the axis servo application manual and elementary schematics to determine correct system of feedback in use on a machine.)

5.7.3 Motor Maintenance-AC

Upkeep and repair of modern AC motors is minimal, and only the OEM should explain if the factory will support replacing motor bearings or encoders in the field. If a motor is bad, it's often just sent to an authorized motor shop for evaluation. Field replacement parts and procedures are located first (and matched up) before pulling out any wrenches.

Just opening the cover to look at the newest encoders is enough to wreck the whole thing, therefore they are best left alone. Among the plethora of alarms captured on the new machines are specific tests for each feedback signal. An encoder accidentally broken on service is quickly detected.

Shocks to the motor shaft can ruin the sensitive encoders. Temporarily removing a good motor from the machine and putting it back on is enough impact to destroy the encoder. Motors are often buried in other mechanicals, or have hard-to-decipher mechanical couplings. Getting help and assistance from the machine builder is the standard protocol for changing out axis motors.

CNC QuizBox

5.2 Do some digging in the references and find ten equations that apply to:
   a.) A machine's mechanical system.
   b.) A machine's electrical system.

5.4 Refer to figure 5.2 and answer the following questions.
   a.) What is the linear distance of the feedback resolution?
   b.) What is the linear distance of one "friendly" pulse?
   c.) What is the linear distance of one complete motor turn?
   d.) What is the expected error pulse at 25% rapid?

5.5 Sketch the D/A signatures for the servo-loop actions below:
   a.) Rapid move at 1000 ipm.
   b.) Feed move at .02 in/rev with a 5000 rpm spindle
   c.) Feed move at 200 ipm.
   d.) A heavily loaded servo system
   e.) An overloaded servo system.

6 *Spindle Loop*

6.1 Introduction

The most powerful motor found on a lathe or machining center is the spindle motor. While axis motors position the tables for cutting, the spindle motor has the serious power needed to make the chips fly. The spindle system must satisfy the computer demands for different spindle speeds.

The spindle drives actively convert the main incoming sixty-cycle power into a carefully controlled variable-speed motor current. Variable speed and optimal torque and power is provided throughout the rpm range of the motor. The primary objective of all drives is controlling the spindle speed during the event of changing load conditions. When necessary, additional spindle electronics are included to perform a limited position mode called orientation.

Spindle *drive* units are the largest electronic part found in the machine and the spindle *motor* is the largest and heaviest motor. Neither unit is easily replaced. When the compact axis servo-units are suspect, the entire unit is simply removed and replaced. But due to their size, spindle drives are often diagnosed at the machine and if possible, internally repaired.

Occasionally the entire drive is pulled out and replaced, usually after repeated field repair attempts fail to find the problem. If local service is unavailable, the entire drive must be shipped to a repair center's test bench. Pulling motors is unpleasant—everything else is first checked, holding motor replacement as a last resort.

Internal repair techniques for machine tool spindle drives is an interesting subject, but clearly beyond the scope of this general text. This chapter will instead concentrate on the features common to DC and AC spindle systems with attention given to those service choices offering the best repair results. Plenty of qualified people in the network can provide free assistance over the phone or schedule on-site internal drive repairs.

6.2   Overview

Spindle motors are the workhorses for machine tools. They do everything from rotating 500 lb. heat treated, cast iron bombs, to milling bury-the-load meter, deep steel pockets. One thing lathes and mills have in common is a hefty ol' spindle motor.

Older turning centers turn gigantic DC motors into step-down belts and stout transmissions to power out long steel ribbons for years on end. New lathes streak around, spitting out another part every ten seconds, the AC spindle load meter a constant blur across the display.

Old spindles come up to speed with a long steady hum, the speed meter climbing and climbing. New spindles emit a brief groan as the speed needle jumps to speed. An occasional click from the motor winding transmission may be heard.

Turning three or four thousand rpm was considered fast for the old spindles. Now with optional spindle packages speeds run fifty thousand rpm and higher. These higher speeds accommodate special cutting applications like fine, ultra smooth capillaries in hard clear plastics, or applied by the large aerospace companies to quickly mill aluminum shapes in a blasting shower of metal chips.

The high speed spindles are run with specially-built motors having ceramic end bearings. Everything is carefully balanced[1] from rotor shaft down to the expensive balanced tool holders. Better finish and feed-rates drive the move to higher speed, expensive spindles. Everyone pays attention when these puppies start winding up.

---

1   An interesting new technology for balancing a running spindle is given by Chris Koepfer, "Tool Balancing on the Fly—It's Coming," Modern Machine Shop, October 1996, 54-60.

More common spindle speeds are somewhere between five and ten thousand rpm, the lathes toward the lower side and the mills on the higher side of the range.

6.3   System Comparison-AC vs. DC

Comparison of the old-DC and new-AC servo systems in the last chapter pointed out important similarities among the axes control systems. This time, the opposite approach is taken to highlight practical differences between the old and new systems for spindles.

Marketing the first machines demanded the best commercial DC spindle technology of the day. Big, two-winding spindle motors offered a viable motion control solution.

The two motor windings are called the *armature* and *field*. By separately controlling, or exciting the field voltage and armature current, the motor responds admirably to the tasks confronting a machine tool spindle. The technology did the job, but technology advanced and a better solution to the motion control problem was presented.

The use of AC induction motors in factories began long ago with applications running two speeds, forward and reverse. These three-phase induction motors ran directly off the power line at a single speed determined by the line frequency and motor pole construction (No=60Fo/P rpm). These old AC motors sorely needed variable speed control like the dominant DC motors of the time.

Then, a breakthrough advance in power semi-conductors fueled new systems capable of variable speed AC motor control. These new drive units arrived on the scene taunting a new control style called *vector* control. To fully understand a vector requires some college level math; but to summarize, the current going into the AC motor carries "vector" qualities that determine motor speed and torque. (Vectors are mathematical constructs to assign magnitude and direction to dynamic quantities. In the case of "AC Vector Control," the current components of magnitude and phase angle are separately assigned and controlled electronically.)

Introduction of AC motors made the expensive brushes and commutators of the DC motors obsolete. By doing away with the brushes, clips and springs, the new simplified rotor designs ran at higher speeds and offered a lower-cost production to factories. AC motors also proved to be lighter and smaller for similar power output, which is a win-win situation for the entire machine tool industry.

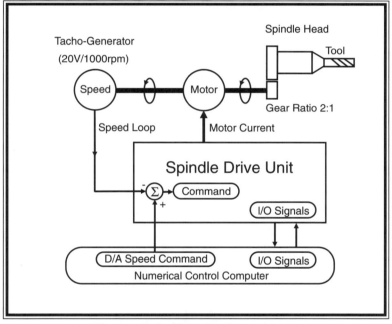

Figure 6.1 Spindle Speed Loop

6.4   Features of the DC Drives

Reliability is the outstanding feature of the trusty DC spindle drives. Once the drive and motor characteristics are factory-matched, the spindle system becomes a tough performer. Even the lightning strikes and power failures seldom result in immediate service problems for these old workhorses.

117

These units use tough thyrister modules to chop up the AC incoming power into the components needed to run the motor. The correct field voltage and armature current is sent to the motor over two pairs of wire cable. Motor power is sure and strong with no hesitation of action.

Spindle speed is summoned by the NC computer using command signals and the release of rotation interlocks. The desired motor speed is actually created and accurately maintained by the drive circuitry and speed feedback from an oversized DC generator tied to the motor shaft. If trouble occurs, protective safety features shut the system down. The complete drive systems are self-contained and, if so desired, yanked out and put on a pallet for shipment to some far away test bench.

Components frequently checked and replaced in the field are discussed in the following sections. This condensed coverage is by no means comprehensive of all the different functions of DC spindle drives out there.

The OEM is always there for the details after an area of discussion is established.

6.4.1   Control Circuit

The old drive's control function is made of friendly, outgoing analog circuits with the related mazes of check pins and potentiometers. The beginning evidence of hybrid circuitry can be seen in the bulging, black, oddly shaped packages found on the PC-Boards. The different layers of hinged boards and chassis are tied together by signal wires all securely screwed into place.

Front PC-Boards called *control cards* talk to the NC computer and synthesize the firing signals sent to the drive's main current generating circuits. These boards carefully orchestrate and literally create the correct motor firing signals for every running condition. During normal running, these firing signals trigger the huge motor currents in the *main circuit*, and the motor responds in kind with ample speed and torque.

The control circuit manages the I/O-Signals between the spindle drive and the NC computer. A few control signals are so common they merit a brief mention. A spindle drive informs the motor status using signals for "motor stopped" and "motor turning." These two signals called "zero speed" (ZSP) and "speed agree" (SPA) reflect the motor action. As usual, the machine's elementary schematics and signal lists will locate specific applications and the adjusted signal meanings.

To put a spindle loop in action, the computer sends out a spindle D/A-command and the motor starts turning. Until the motor climbs up to commanded speed, the entire system sits and waits—waiting, waiting, waiting for the *SPA* signal. Once correct rotation speed is acknowledged, the computer can move on with its agenda. From running speed, a spindle stop is acknowledged by the *ZSP* signal. After the stop *command*, the system again waits for an indication of zero speed and keeps waiting until it arrives. If either of these signals are faulty, the machine hangs up, waiting for the status acknowledgments that never come. Often small reed relays mounted on the control card are finally responsible for sending out a drive's status signals.

Alarms on either the spindle or NC-side will interlock the spindle and keep it from running until the cause is identified and cleared.

Providing smooth, safe, variable-speed spindle power is a glowing testament to those skillful analog circuit designers of the time.

6.4.2   Main Circuit   *Older Systems*

The spindle drive modulates a fixed frequency AC source into the dynamic DC current for running the motor. This conversion of source current to motor current takes place in an area defined loosely as the *main circuit*, which safely channels the heavy drive currents through power semi-conductor modules mounted deep inside the drive chassis.

Currents and voltages inside the main circuit are very dangerous—remember, these switching currents are powerful enough to run a massive spindle motor.

If the main circuit is de-energized while the the control boards are left running, the logic level firing signals remain active, except that no large motor current can result. This condition is called *control power only* status. The main circuit is off, but the drive's logic control is still running.

Factory-authorized repair centers test defective drives by connecting them to to simulators in this "control power only" status. It's safer to check the firing signals and internal PC board operations without all those big currents flowing around. After all the blown circuits and intermittent problems are safely identified and the defective board-level and chassis-level parts replaced, the main circuit is gradually brought back to life to finish testing the refurbished unit.

6.4.3   Alarm Detection   *Older Systems*

Front line protection for the drives comes from *quick-blow* fuses on the main input and motor current output. These fuses blow for a reason—either the incoming power source, drive unit, or motor has a problem. A blown fuse triggers an alarm switch contact which stops the entire machine.

The drive unit monitors itself for problems and also listens to the NC computer for evidence of alarms elsewhere on the machine. When an alarm is caught by the drive, the motor current is immediately blocked for safety. Shorts in the main circuit, mixed-up control signals, as well as blown fuses and heavy cutting, send up alarms. The entire machine reacts to spindle alarms in a way specified in the sequence software written at the factory. This is how alarms are handled, not just spindle alarms but all types of alarms. The black-and-white alarm sequences are supplied in the PC ladder diagram.

Alarm lists and maintenance procedures are sparse for some older DC drives. Any notice of alarm lights on the drive help diagnose the problem.

6.4.4   Run Enable

For a lathe, the spinning parts must be firmly clamped and the safety door to the machine closed before the spindle will turn. This check of safety features is done by the NC computer.

The sequence ladder determines the conditions that interlock spindle rotation. When all the safety interlocks are approved, the spindle drive unit will run a programmed spindle command. The builder's advice is needed when a spindle *run enable* hangs up.

6.4.5   Replacement

An experienced CNC serviceman who specializes in the internal field repairs of old DC spindle drives carefully collects data on the original system before swapping out any PC-Boards. All the defective thyrister modules and PC-Boards are identified and replaced at the same time.

Replacing the individual control cards on the drive only solves the problems handled by that board. Many times the damage to PC-Boards occurs when a connected module shorts. Installing a new board into this shorted circuit will only damage the new board and cost a few thousand dollars.

Whoever does the work will make nice drawings of all the original wiring to the drive and miscellaneous part layouts before any disassembly begins. Every potentiometer and jumper setting is written down in a tablet using detailed drawings. The OEM will fax basic drive checking procedures to experienced maintenance people.

The entire spindle drive has a main model type and power rating. All the printed circuit boards used in the drive also have model numbers. To order up the correct spindle drive version the supplier is called with this information. The supplier, which usually is the OEM, will recommend the best way to remove the spindle drive from the machine.

Extreme caution is followed in the internal repair of spindle drives. A qualified repairmen will disable the heavy current/voltage section of the drive and perform adjustments and testing with the power removed. Spindle repair data is collected using detailed investigations and hard-earned experience to minimize the loss of expensive boards and modules. They can best decide if a complete drive replacement would actually end up being the cheapest repair route.

6.4.6   Specifications

Older Systems

Spindle output power is carefully detailed in the machine specifications. The turning strength of a spindle motor is an important characteristic for moving production parts off the machine. Overly heavy cutting or other excessive demands will shut down an under-rated spindle system.

A desirable feature for all spindle systems is constant motor power over the largest possible band of speeds. DC spindle drives output full power after reaching *base speed*; once above base speed, the drive power remains optimal.

Mechanical transmissions increase the range of constant-power. By gearing up the spindle motor into the constant power regime a low speed spindle operation can maintain optimal motor power. Accidentally running a low gear application in high gear is a common mistake heard in the field.

6.4.7   Orientation

Older Systems

When the spindle motor shaft needs to be stopped in a certain alignment (for operations like tool changing), a spindle drive will switch into a limited positioning mode called *orientation*. Spindle alignment in program will use the command for orientation (often the M-code assigned is M19).

During the brief spindle positioning process, a position sensor indirectly commands the drive unit. As the spindle gets close to the alignment position, the drive issues *proportional* commands until the motor hunts down the proper alignment position.

Orientation on old DC spindle systems is a real stinker to adjust. The system works great on machines with low hours, but after the machine ages fifteen years, the combination of mechanical gear backlash and drifting analog circuits send the orientation adjustments into the halls of some unlucky factory representative. These are the tough service calls that end up on factory service schedules.

6.5   Features of the AC Drives

Spindle control technology of AC motors has departed from the DC systems of the past. New spindle designs combine advanced control schemes with power semi-conductor electronics to exploit the benefits offered by the compact, low-cost, low maintenance induction motors.

These "high technology" drives have a *converter* section to change the AC input power into a stable (300 Vdc) power source. Then, an *inverter* section, which uses power drive transistors and *vector* current control schemes, chops this source into a new three-phase AC current for running the spindle.

The transition from mixed (analog/digital) designs to fully digital control expanded both the flexibility and operating characteristics of these AC drives. Now the same drive is applied to different motor types by simply matching software routines and settings for the new motor characteristics and application preferences.

During the development of these drives popular styles for drive transistors have evolved from the bipolar junction "hockey puck" transistors (BJT) to the modern IGBT modules of today. Transistor improvements give better (static and dynamic) switching characteristics that led to better approximations (chopping) of the desired AC motor current.

123

Circuits for detecting motor speed and current feedback circuits also improved. Early speed detection came from AC signal resolvers, later updated to pulse generating disk encoders. Current detection schemes started with isolated shunt resister circuits, since replaced by nifty coil wound current transformers with built-in signal control amplifiers.

Documentation has steadily improved as the units improved. New machines receive a hefty spindle maintenance manual as part of the basic documentation.

6.5.1   Control Circuit

Newer Systems

Carefully timed firing signals for the drive's transistorized inverter section are sent from the logic control circuits. Firing signals are likewise provided for regenerative braking and converter section control (if the converter uses active rectifier control).

The control section also provides detection for drive unit alarms and I/O communication. Finally, the control circuitry monitors the motor current and speed feedback to achieve the listed drive objectives.

Plugged into the drive chassis are the control printed circuit boards. Two central boards are called the control card and base card. Other optional cards are fitted to enable spindle orientation, motor changes and other OEM options.

6.5.2   Main Circuit

Newer Systems

Everywhere the heavy motor current travels is considered a part of the main circuit—all the high current modules involved in building the *converter* and *inverter* sections are involved. Regenerative braking transistors and large bus capacitors are also included. A main circuit diagram is either provided in the drive schematics or available from the OEMs.

The main circuit is closely monitored by the control circuit logic. Heavy current components going out to the motor and between drive stages affect the control firing signals. The roughly 300 Vdc bus voltage will activate the regenerative firing signals if it climbs too high during a motor deceleration. Newer drive types control the initial build-up of bus voltage to protect the new technology, IGBT type transistor modules.

When a semi-conductor fails inside the main circuit, the neighboring modules may also receive damaging levels of current. The short may also damage the firing signals sent from the control amplifier circuit causing additional, expensive damage to the drive.

6.5.3 Alarm Detection

Older style AC drives, with mixed (analog/digital) circuitry, indicate alarm conditions by lighting up LEDs on the control cards. These single spot LEDs were subsequently replaced by digital seven-segment alarm displays. Finally, key-operated, multi-functioned display monitors appeared on the "mostly, if not all" digital drives.

The meaning of a spindle alarm is given inside the drive maintenance manual.

6.5.4 Run Enable

Conditions for a spindle rotation are met before the motor will start turning. Newer units, with active converter sections can automatically lower the DC bus voltage, making motor rotation impossible until the run enable interlocks are approved. Without adequate bus power, motor current is not possible.

All the drives require signals for speed command, run enable and direction before the motor turns. Some units have small override control panels mounted on the drive for testing the spindle.

6.5.5 Replacement

When ordering a new digital drive unit, the factory will modify software and load up parameters, so it will "Plug and Play" like a new modem. At the time of a parts request, the exchange procedure is reviewed over the phone, just in case the thing plugs but **doesn't** play.

Software-controlled digital units need the original parameter data, correct software chips and correct hardware versions.

Mixed (analog/digital) units require the adjustment of several jumpers and potentiometers. The factory procedures for a specific drive replacement provides the necessary replacement information.

(Complete drive unit replacement is primarily guided by OEM supplied procedures and suggestions. Further guidelines were reviewed in section 6.4.5.)

6.6   Spindle Motors-DC

Down on hands and knees, looking in on a 500-pound spindle motor tucked up within four surrounding walls of machine casting tends to discourage the opinion of a bad spindle motor. To pull an old DC spindle motor from a lathe is a heavy, dirty, thankless job. The motor is securely bolted up inside the machine casting to feet molded on all four corners of the motor frame.

The foot mounted motor sits horizontally inside a lathe. Flange mounted motors go up vertically on the head of a machining center. The motor frame evenly distributes the motor weight into the main machine casting.

As expected, a big DC spindle motor has two main parts: a big spinning part and a big stationary part. The spinning part is supported in front and back by high class bearings. The motor shaft pokes out the front and the "TG" speed sensor hooks to the back. Armature windings and brush commutator are mounted somewhere in between.

Long, soft carbon brushes ride on the commutator to provide powerful currents to the spinning armature. The fixed chassis of the motor is interwoven with the long wire coils of the field windings. The field winding supplies an attractive magnetic field which the armature tirelessly pursues.

All the surging currents within the motor windings leave behind heat. Around the perimeter of the motor frame are ribbed cooling fins molded in the casting. Electric fans mounted to a sheet metal cooling shroud force cool air across the warm motor fins.

6.6.1   Specifications

Older Systems

Riveted to the motor frame is the name plate that contains the basic electrical information of the motor design. A motor shop relies on this plate information to order parts and accurately test the motor under load before shipping it back to the customer.

A typical name plate has the benchmark values for the field and armature circuits, along with the power ratings and speed characteristics of the motor. The OEM can provide other motor specifications if service to the spindle motor becomes necessary.

Nominal values of resistance for the field and armature windings come from the factory. Values are different for "motor hot" and "motor cold." When a field rewinding is required, maintain the wire length and diameter used in the factory's original field windings. In this way the new windings will best reproduce the original magnetic and resistance properties designed by the factory

Other useful motor specifications are diagrams showing a motor's voltage, current and torque as a function of speed. These dynamic characteristics verify a repaired spindle system is back to factory original condition.

Additional motor questions sent to the factory include the physical motor dimensions, brush parts and any newly developed retrofitting options for vintage spindles.

6.6.2   Feedback

Accuracy controlling the speed of an industrial DC spindle motor relies on the accuracy of the feedback detection. Dependable speed feedback leads the quest for smooth, powerful motor currents and tight control of the motor speed.

A multi-brush tacho-generator mounted to the end of the motor, opposite the drive shaft, outputs an increasing DC voltage as the speed climbs. This voltage is used at the drive to maintain correct motor speed. If the motor bogs down, the drive sends more current; if the motor speeds up, the drive reduces the motor current. This is a closed speed loop in action.

After years of faithful service, the generator develops deep grooves from brush wear. If the grooves are small, a good cleaning and re-surfacing may extend the life of the unit. Complete replacement is possible after the speed signal becomes irreparably degraded.

As usual, procedures and parts come from the OEMs. Care is required during assembly because the strong permanent magnets used around the generator can suddenly pull in on the brush holder, shearing off the soft, tiny brushes.

Polarity of the TG signal is critical; a miswiring causes a dangerous and frightening maximum speed runaway of the spindle motor. The drive keeps sending more current because the motor keeps going faster in the wrong direction!

6.6.3   Motor Maintenance-DC

When a motor burns up, it's pulled out and sent to a qualified motor shop. After the motor receives a thorough cleaning, it is deemed repairable, or unrepairable and in need of replacement.

Simple preventative measures for maintaining spindle motors can save a company money. The factories offer maintenance schedules for checking the main motor brushes and cleaning the motor cooling fans and dirt screens. The factory can send out the latest procedures for inspecting and replacing a specific set of spindle motor brushes.[2]

The life of the motor is extended with periodic cleaning and fresh brushes, but as pointed out, there are risks. Before checking or cleaning an antique motor the replacements are made available, just in case it should mysteriously short out. Cleaning motors is a big responsibility; if something goes wrong, an old machine will stay down for months waiting for a rare replacement.

Care is taken to keep foreign agents from contaminating the inner cavity of the motor. Any evidence of arcing in the brush area is critical. The commutator should look light brown, but not black. A burning aroma or damaged commutator bar indicates an expensive problem.

The brushes are carefully removed after noting the electrical connection of the wire pig tails for each brush. Motor brushes are always handled with clean, dry hands. When symptoms of wear to the brush or pig tail are found, the brushes are replaced and the loose carbon dust is thoroughly blown out with clean, dry, compressed air. The new motor brushes are reinstalled and the coil spring brush clips properly seated.

The face of every brush must ride evenly on the commutator. All screws and covers are replaced, and the assembly notes double checked to make sure everything is back to original. A gentle spin of the motor is done after everything is assembled. Initially the motor is run at its slowest speed in both directions. If an alarm occurs, all the service steps are carefully retraced to find the error. After the brushes become seated, the spindle speeds are gradually stepped up. Spending an hour or longer to gradually break in a new set of brushes is time well spent by the service engineer.

2   See the reference given in Section 5.6.3

Complete motor removal takes the big wrenches and an overhead crane or hoist. The machine builder's guidance and assistance interpreting the machine drawings should set the course of action long before any motor connections are loosened. This is a big job for a service engineer. During a complete removal, the motor wire connections are unhooked inside the electrical cabinet, or at the motor junction box. Wires are never cut.

Local motor shops will clean and evaluate a DC spindle motor. When burned armature wiring is found sealed inside epoxy casts the entire armature is replaced. As always, hidden dangers are present. Factory-authorized shops know the hidden dangers because they routinely replace complete armatures, end bearings, tacho-generators and field windings.

A spindle motor shipment requires a few days for a truck carrier to provide surface shipping of these heavy motors. They are tied secure to a heavy crate or pallet for the journey.

6.7   Spindle Motors-AC   *Newer Systems*

Industrial grade induction motors for machine tools offer both power and speed in a compact, efficient package. Compare the size of a 15 kW DC spindle motor with the like-powered AC motor—it's about half the size. The comparison further improves with newer style motors having the liquid-cooled jackets.

As expected, the AC motor has a spinning part and a stationary part—and that's about it. The rotor spins on two high class bearings mounted inside the stator. The motor's drive shaft runs out one end, the motion encoder out the other.

With no brushes or commutator to worry about, the simple rotor can spin much faster. Heat build-up is reduced by forced air ventilation, frame cooling fins and/or hollow liquid filled rotors and motor frames.

The motor cavity is kept sealed from the shop environment. The sensitive encoder is also mounted in a clean and sealed section of the motor. Oil seals on the front shaft and electrical "cannon plug" or junction box connections in the rear help keep the motor dry. However, they are not rated for submarines; poorly located motors will have problems if fluids constantly splash or drip on the motor.

6.7.1   Specifications

The variable speed AC motor is a newer technology, and the factories seem to provide better documentation for the many electrical and mechanical specifications.

Characteristics, vibration standards and cooling systems are all detailed in the spindle drive maintenance/application manuals. Necessary values for calculating a spindle "duty cycle" are listed. Other motor questions should forward on to the factories.

6.7.2   Feedback

Early AC motors used sturdy, dependable resolver feedback units. These popular units were gradually replaced by the disk style, signal pulse feedback units.

Speed resolvers have two stator windings placed at right angles to the motor rotation. Constant frequency signals excite the stator windings. The rotor motion then excites another winding with a signal frequency related to the rotational speed of the motor. The spindle drive can then determine the motor speed from this resolver signal. Motor shops routinely replace damaged resolvers. (How the stationary resolver windings are initially damaged is hard to imagine.)

Magnetic pulse encoder style feedback units act like cassette tape players. A tape pickup head reads a spinning magnetic coating painted on an aluminum disk. The high performance, digitally controlled spindle drive count on these encoder pulses to decode, among other things, the motor speed information.

When a motor encoder fails, the motor manufacturer advises if replacing the encoder is allowed. Replacing an encoder on a motor having excessive bearing run-out leads to a very short life of operation. Generally, a complete motor replacement is advised. Sending a faulty motor to a factory authorized repair center offers a better environment for encoder replacements, motor testing and factory adjustment. Considering how tough it is replacing a spindle motor, the last thing anyone wants to do is go back and replace it a few weeks later.

6.7.3   Motor Maintenance-AC

These motors are touted as maintenance free in the sales brochures, but occasional maintenance does takes place.

Shroud-mounted cooling fans often get plugged up with gook. Once they seize up and stop spinning, the fan burns out. Digging out the grimy, burned-out fan is then required. Out in the open this is no problem, but otherwise, a complete spindle motor removal may be needed. Fans can be kept clean with an installation of high-flow disposable furnace filter material. As the material darkens from the gook, it's time to peel it off and press on a fresh piece.

Some motor applications require specially-built factory wrenches to pull the motor away from the machine. The machine builder is called for advice and assistance. Likewise, if a motor starts suddenly getting louder, the factory is kept abreast of the situation before the machine goes completely down. Spindle motors are generally stocked at very low levels because of their overall reliability.

CNC QuizBox

6.4.1 Draw a circuit having a 220 Vac 3ϕ source that demonstrates:
a.) Thyrister rectification.
b.) Transistor rectification.
c.) Diode rectification.
d.) Calculate the maximum level of rectified DC voltage.

6.4.2 Sketch a simple block drawing of a DC spindle drive showing the :
a.) Main Circuit for the Armature.
b.) Main Circuit for the Field.
c.) Control Circuit showing the command and feedback speed.
d.) Control Circuit showing the firing circuit.

6.5 Sketch a simple block diagram of the following circuits used in an AC spindle drive.
a.) A passive converter section.
b.) An active converter section.
c.) Inverter
d.) Control and base firing signals

6.6 For a DC spindle motor:
a.) Describe an equivalent circuit model for the motor.
b.) Sketch the armature voltage and field current across a range of typical motor rpm.

6.7 For an AC motor:
a.) Describe an equivalent circuit model for the motor.
b.) What are the normal resistance readings for the terminals of a 3ϕ squirrel cage induction motor?

7 Control Features

7.1   Introduction

The basic machine systems introduced in the last four chapters are only components of the overall machine. Now it's time to start collecting everything together and consider the overall function of CNC machine tools. In this chapter we explore the operation and repair of machine tools from a control standpoint. In upcoming chapters we shift our focus to include mechanical systems and the availability of optional equipment.

7.2   Zero Return

First thing in the morning, the operator powers up the machine and does a *zero-return* (ZRTN). This action, also known as zeroing-out of the machine, is required (see note below) before running a part program. After each axis is selectively moved to its physical zero spot, a return light comes on. A manual or automatic system continues this zero-return operations until every one of the machine axes finds its physical *ZRTN* position.

Once zeroed, the machine remembers the set positions, until the next time the machine is powered down.

(Note: This is true unless the machine has absolute positioning systems, in which case the zero-point should never get lost or require initialization.)

Two important CNC functions occur when those zero-return LEDs light up: the position displays update to reflect the new machine zero and a set of safe motion limits are calculated using the new zero point information.

Programmed position commands for a machine are always made in respect to this fixed zero reference position. While the part programming itself is done in a convenient *coordinate system*, all systems are somehow geometrically linked to the first simple system established by the zero-return. Good programmers will quickly explain the coordinate systems they are using in a program.

Soft over-travel (soft OT) limits are the safe motion limits picked up at *ZRTN*. These limits are called "soft" because they are stored in software memory. They are "over-travel" because should the machine try to cross one of these software controlled boundaries, an alarm is issued, stopping the machine dead in its tracks. It is important to realize that on incremental machine tools no soft-limits exist prior to zero-return. Only the hard limits will stop an end travel before a *ZRTN*.

Observing live action on the machine during a zero-return illustrates a subtle sequence of events. To begin, an axis is moving at a rapid rate towards the zero-return point. Suddenly, the motion slows to a creeping speed and continues until the return light blinks on. When everything is working the same spot is found every time.

This sounds simple enough, but behind the scenes, several key actions are involved. The initial slow-down, or deceleration, was caused by a physical limit switch mounted on the machine, appropriately called the *decel* limit switch. After receiving the decel switch signal, the control begins looking for a marker pulse in the motor feedback signals. A *marker pulse* is just a narrow electrical pulse, sent only once during every rotation of the motor.

Once the marker pulse is received, the motor is slowed a second time, allowing for an exact countdown of the positioning type feedback pulses. A zero-return *register* on the control is counted down by motor position pulses. When the register count is zero, the *ZRTN* light comes on. The number of pulses (i.e., distance) from the marker pulse to the *ZRTN* position is adjusted by the *stroke*, or traverse distance parameters assigned for each axis.

7.3   Safety Interlocks

A machine operator is responsible for the safe operation of a CNC machine. With an errant push of a button, a half ton column can accidentally *crash* into the support table. When accidents of this nature happen, the resulting machine damage can take weeks of costly mechanical and electrical repairs.

While working around $300,000 dollar machines, the operation is best left to the owner's trusted on-site personnel. The event is quickly forgotten if they accidentally crash it down, however, a visiting serviceman's crash goes into the annals of CNC history. Experienced CNC operators run these machines for years without allowing a machine crash. A visiting serviceman may never have moved, or even seen such a machine.

Certain emergency situations were anticipated by the machine builder. To counter these dangers, the machines have numerous factory installed safety interlocks. Some interlocks are activated when machine conditions or user settings are improper for the actions attempted. For example, trying to run the spindle without proper clamping of the spindle chuck is not allowed. The spindle will not be allowed to run until the condition of the chuck is corrected.

If during the operation of an automatic parts program the cycle button suddenly stops executing the program a safety interlock has been violated. This "cycle-start" interlock has several possibilities. The safety door may not be closed tightly, the status of an external device, like a bar feeder may have a problem, or possibly the lube oil for the machine is low. Until the exact reason is found and remedied, the machine won't go in auto.

The tool changing carousels have a plethora of possible interlocks. Once activated, these tool changer interlocks are notoriously pesky to resolve. The use of a pallet changer brings in another list of safety interlocks. Usually these interlocks are familiar to the experienced on-site operators.

Trying to reset obscure interlocks on a tool carousel or pallet changer is tough, a cooperative operator may cheerfully clear up these kind of troubles in a few minutes using their home-brewed, secret recipe sheets. When nobody is familiar with an interlock, the service engineer will have to settle in for the long haul, checking sequence addresses and switch diagnostics.

Machine interlocks perform absolutely critical functions that are not always fully appreciated. It's very difficult to comprehend all the critical safety functions the sequence interlocks perform. To be safe, never allow anyone to apply CNC knowledge to defeat these vital systems. Here the popular saying, "A little knowledge is dangerous," aptly applies. The interlocks must be kept in factory original condition. This requires that everyone's "proposed" service solution stay within the factory designed interlock boundaries.

7.4   Over Travel

The axis motors are mechanically coupled to long threaded ball screws, securely supported on both ends by heavy duty sealed bearings. Together, the entire motor shaft and screw assembly rotates forward and in reverse. A large nut riding on the ball screw is screwed in and out. Tied to this motor-driven ball screw nut are the big, movable machine castings which slide on smooth lubricated *guide-ways*.

In this way, the rotating axis motors can smoothly move the big machine castings back and forth. The type of castings depend on the type of machine. In the case of a simple three-axis mill, the castings include a two-axis work table with side-to-side and front-to-back motion and an overhead column with an up-and-down motion. In the case of a simple two-axis turning center, the complete tool turrent casting moves up-and-down and side-to-side.

In these simple layouts, the maximum travel is obviously limited to the length of the ball screw. Moving beyond the end-distance would jam the nut and cause serious machine damage. At each end of the ball screw is a collection of over travel switches designed to safely contain the machine's travel. If a casting moves too far down on one end, a switch trips, sending instructions to stop any further motion.

End travel switches are called the *hard* over travel (hard OT) limit switches. When the machine lands on the hard switches, the motion stops. Sometimes, special procedures from the OEM or a friendly on-site operator are needed to get it back off.

After zero return, the soft over travel system takes over. When the machine is zeroed-out, the software travel limits, or stored strokes, are established. *Stored stroke* limits set a boundary of allowable moves. Trying to move the machine across these boundaries is prohibited. Interestingly, the NC computer knows the line of these boundaries by comparing the current position register with the known limit in that direction (calculated by adding the stroke parameter value to the known zero point). When the numbers come up equal, the computer shuts off the move command and issues a *soft OT* alarm. Simply moving back away from the violated soft boundary releases the stored alarm condition.

7.5   Circle Cutting

Cutting a circular arc tests the related motion of two independent axes. It's just the nature of a circular move. The simultaneous and synchronized movements of two right-angled axis motors must be perfect, or the shape and finish is not a correct circle. A machine cutting good circles is a good machine on many levels.

Circle cutting adjustments for CNCs run the gamut from control side compensations to mechanical-based alignments. For the case of oddly shaped, non-symmetrical bulging circles a lengthy machine-side investigation is often indicated. Control-side techniques include making adjustments to the backlash compensation parameters to smooth over the steps left at the circle quadrants, or tuning the unbalanced servo loops that leave circles elongated and egg-shaped along the diagonal axes.[1]

During a full circle move, the *speed* of the two opposed servo loops becomes equal four times. If these related speeds (at each diagonal point) have drifted, a common set of service adjustments are made to restore the related rates needed for a good circle. Specific procedures for balancing the circle settings depends heavily on the control type. The old analog servo loops often drift and usually need adjustment, especially after analog components in the loop are replaced. Newer digital loops rarely drift and are generally unaffected by part replacements.

Steps at the circle quadrants are removed by adding or subtracting pulses from the *backlash compensation* parameters. If a quadrant step moves in a distance of six pulses on one side of the circle, and moves out a distance of four pulses on the other side of the circle, a good backlash setting would be around five friendly pulses—the average of the two.

Backlash compensation is a standard function for all CNC machines. The more specialized *circle-cutting* compensation systems, available in different formats from all the major control builders, use long lists of complicated factory settings to adjust for better circles. These special compensation systems are set up at the factory for each machine type.

Inspecting the final shape of a circular metal part highlights the mechanical and electrical condition of the machine. Even fine adjustments made to the machine or control is quickly reflected in the final shape of a circular cut.

Better circles are a strong point for competition, so elaborate compensations are worth the expense to develop. The field adjustment of these special factory compensation systems requires OEM experience and the advanced use and interpretation of external test equipment (specifically, multi-channel chart data recorders).

---

1   See Appendix Two for the best procedures to request from your OEM. See also Section 1.13 and Section 10.8.

In Part Two circle cutting is discussed as a diagnostic tool for checking a machine's overall health. Both the mechanical- and control-side systems must be fundamentally sound to cut a good circle.

7.6   Closed Loop

A leap in machine tool technology occurred with the introduction of closed loop motion control. Efficient, closed motion loops arrived with the invention of computers capable of tracking all the loop's static and dynamic characteristics.

The NC parameters are loaded up with these static and dynamic *constants* derived from the machine's actual construction choices. To get an idea of these motion loop constants, return to the example of a simple axis move. Instead of a ten-inch move, imagine a short *metric* move of ten-millimeters, and, to make it easy, assume the pitch of the ball screw is this same ten-millimeter distance. The machine makes the move by turning the motor and ball screw combination one complete revolution.

From this simple example what are the motion loop constants? Well, the pitch of the screw is ten-millimeters. The feedback unit sent a fixed number of pulses during one revolution. There's the unknown free weight of the machine being moved by the ball screw and the overall stiffness of the underlying machine castings. How about the mechanical backlash values and motor size? All these things are examples of loop constants that are calculated and set up in a motion control application.

Now change things around a little: suppose a smaller machine, using a twelve-millimeter screw with a different style feedback, is built and controlled by the same control computer. New loop constants are calculated for the application and then entered into the control computer's setting parameters. The same NC computer can be set up to run different applications.

Calculating motion constants is a job for factory application engineers.

7.7   Control Cabinet

A machine is outfitted with cabinets to protect the electronics and maintain adequate ventilation while staying free of oil, dirt and metal shavings.

Electronics are very sensitive to temperature. To keep things cool, a sealed cabinet will use *heat exchangers*. Heat is transferred out without any dirty air actually entering the machine cabinet. A heat exchanger will transfer excess cabinet heat across a wick using two fans. One fan blows cool, but dirty, shop air across one side of the heat wick while another fan circulates the warm air inside the cabinet across the wick.

High power control components have aluminum heat sinks to help dissipate their excess heat. Air is directly blown across the fins of these heat sinks to efficiently carry off the heat. Heat sinks generally protrude outside the machine cabinet, but inside an enclosed *air box*. This hollow air box sucks cool air up from under the machine, across the heat sinks and out the top.

A rule of thumb for cooling fans is *up and out*. Since heated air likes to rise anyway, the heated air is blown **up**. To keep fans from spitting oil on the cabinet electronics, the clean filtered cabinet air is blown **out** into the dirty shop environment. A properly ventilated cabinet feels cool to the touch, running warm cabinets reduces the life of the electronics.

All ventilation filters should be regularly cleaned. If replacing these filters is easier than cleaning, some Velcro strips can be applied around the air intakes. High-flow white furnace filters are cut to size and slapped in place. When the material becomes discolored, they are pulled off for a fresh piece.

7.8   Relays and Contactors

Controlling a heavy motor current with a small logic level current is accomplished using *magnetic relays.*

Relays come in many shapes and sizes. The smallest are miniature reed relays mounted to printed circuit boards. Another kind of relay most commonly found in machine tools is the *ice cube* relays, their status always on view through their clear plastic covers. Still larger magnetic *contactors* are fitted to switch hundreds of amperes of electric current. Start with a tiny reed relay switching an ice-cube relay which in turn switches on a giant contactor, and it's easy to see how a small signal current is able to switch on just about anything.

Every relay has a coil-side and a contact-side. When the *coil* is energized, the *contacts* close. To operate a relay a small signal is passed through a long coil of fine wire, a strong magnetic pull develops which trips the heavy-duty, counter-balanced relay switch contacts.

Discussion of a specific relay calls attention to a specific coil voltage and contact current. The coil voltage and contact rating is always maintained at the original factory specifications, and if possible, with the same relay maker. Accidentally installing a 24 Vdc ice-cube relay into a 100 Vac socket won't work. A magnetic contacter having a 200 volt coil can't energize on 100 volts. Accidentally connecting a 100 volt coil into a 200 volt circuit will quickly burn up a coil.

The largest relays are referred to as magnetic contactors. They contain many sets of contacts for status, alarm functions and main current switching. When large currents are passing through a contactor, any interruption causes an arc across the main contacts. Pits in the main contacts develop as a result of this electrical arcing. Occasionally, an arc will actually weld the contacts permanently together. The factories try to reduce the "cycle of arcing" in all contactor circuits.

7.9   RS232 Communication

A remote PC computer linked by cable to the main NC computer provides a fast exchange of program, offset and parameter data. The same serial communication link lets a small laptop run a towering CNC machine with a *spoon-fed* part program.

A full discussion of *serial* communication standards is lengthy, and readily found in other books and references.[2] The goal here is to list only the standard terminology and features to recommend a further study. The discussion begins with the current industry standard for data transfer called "serial RS232 communications."

The so called "serial data" is sent in *code*. Codes go back to the time when tele-type machines were all the rage. Someone figured out that with only eight separate *bits* of data, each character on a standard keyboard could be assigned its own distinct data combination. One standard set of combinations is called the ASCII set of character codes. It follows then that every character in a G-code part program is described by a single *byte* (8 bits) of data.

The speed that serial data moves is described by the *baud* rate, given in units of bits-per-second (BPS). The traffic control for all this moving data is defined by *handshaking* agreements established between the sending (PC) and receiving (NC) computers.

The normal handshaking agreement used with NC systems is called software handshaking control. In this handshaking scheme, all the "stop sending" and "start sending" codes are carried right along with the other strings of communication data.

---

2   Industrial Data Communications-2nd Edition, Lawrence M. Thompson, Instrument Society of America, 1997.

Another method less used is called *hardware* (RS/CS, or RTS) handshaking. This scheme adds two more "hard" wires to the communication cable to control the starting and stopping of data traffic between the two computers.

The cable schematics are modified to handle either the software control or hardware control handshaking. For the system choices of handshaking, baud rate and data codes to work, the cable and communications settings must match exactly on both ends of the cable.

When early CNC machines entered the market, a short program was keyed into the small memory and was run directly. Any lengthy part programs that exceeded the small memory size were run remotely off a long paper or foil punch tape. These paper tapes and tape punch systems are still occasionally found in operation today. This concept of loading and running a new machine in "tape" mode survives from the past.

While RS232C (now in the E version) is currently a very popular communication choice, it has performance limitations. Today, really long programs (like a 100 Megabytes) can result from a CAD/CAM generated tool path. This much program data far exceeds proprietary CNC memories, so a higher-speed version of tape mode is required. (Or an altogether different CNC system offering open, robust communication features).[3]

System standards like Ethernet[4] offer the advantage of speed and performance as well as the transfer of more types of data than just part programs. These new communication systems allow high-speed control of external devices like servos and spindles, or linking together entire groups of machines. Follow up material is available in the listed references.

---

3   In 1997 the US auto industry issued a call for "Open Modular Architecture
    Controllers (OMAC), on the web at www.arcweb.com See also Section 9.2.3.
4   Golden E. Herrin, CIM Perspectives, "Ethernet Flourishes In the Shop," Modern
    Machine Shop, March 1996, pg 158-160.

CNC QuizBox

7.2.1 Sketch the axis speed during a zero-return sequence, for a machine tool outfitted with a deceleration dog switch. Include the switch signal, marker pulse and stroke distance in the drawing.
   a.) Repeat the sketch for a high-speed memory zero-return.
   b.) What is the distance from zero return to the marker pulse?
   c.) How can you calculate the distance from zero return to the dog?

7.2.2 To increase the zero-return stroke by 0.0320 inches requires how many 0.001 mm parameter units?

7.3 List some reasonable interlock conditions for the following:
   a.) Spindle Start.
   b.) Cycle Start.
   c.) Feed Hold.
   d.) Emergency Stop.

7.5 Sketch a circle having the following machine tool problems:
   a.) Excessive mechanical backlash.
   b.) Excessive cutting rate.
   c.) Unbalanced servo position loops.

7.6 Using the results of Exercise 5.2, list the specifications required to completely describe a:
   a.) Two axis lathe system.
   b.) An add on rotary axis system.

7.9 Sketch the schematics for a new RS232 cable having a 9-pin connector on one end and a 25-pin on the other for:
   a.) Software handshaking.
   b.) Hardware handshaking.

8 *Mechanical Features*

8.1   Introduction

Backbone mechanical components—built by the machine maker—provide the working foundation for control electronics. With a good cast foundation, anything is possible. But if the castings give, no amount of electronic trickery will work. Machine- and control-side repairs are often dealt with separately, so when trouble, or just good questions, arise about a machine's mechanical systems, the machine maker has the responsibility to know—or to find out.

Machine Side
Guy

8.2   Ball Screw

A ball screw converts a rotary motion into a linear move. The free length, diameter and pitch of a ball screw is selected by the machine builder. Free length of the screw determines the stroke of the machine tool; the pitch determines the distance covered per revolution of the electric motor and related accuracy-versus-speed issues; and the screw diameter satisfies the load requirement of the system.

Using heavy support bearings on both ends, the ball screw is held rigid within the machine casting. A motor is then coupled to one end of the screw, so as the motor rotates, the screw rotates. The rotation transfers its motion to the nut that is slid up-and-down the length of the screw. Connected to this nut are the actual machine fixtures.

Ball screws are supported in the machine by thrust and load bearings in different design combinations. Techniques to pre-load the ball screws are practiced by some machine builders. If a ball screw support becomes worn, machine accuracy suffers. The screw backlash is checked with a dial gauge against the screw-end face. If the face is moving excessively in the bearing support, the system needs a mechanical service.

The distance of slide-per-revolution is precisely determined by a thread pitch that is factory ground into the ball screw. A common pitch for a metric machine tool is ten-millimeters. During the production of a ball screw, extreme care is taken to maintain an accurate pitch along the length of the screw. The metallurgical composition of the ball screw controls both the thermal expansion and overall wear qualities, two important concerns in a machine tool application.

A tight-fitting, well-lubricated, quality ball screw is required if an electric motor is expected to drive the machine back and forth in a controlled, repeatable fashion. A specifically formulated "Way Lube" oil is pumped out to the ball screw and ways to keep things running smoothly. Occasionally, extra grease Zerk fittings are added to augment the lubrication.

With the advent of linear induction motors, the ball screw, bearings and couplings are gone. Only the table fixtures and guides remain.[1]

8.3   Guides and Ways

The machine guides come in a spectrum of shapes starting with the box-type way common on older machines and ending with flat linear-ways found on the fastest new machines. Guide-ways or *guides* and *ways* are treated in the references.[2]

In between box and linear are a wide assortment of styles looking for the best engineering solution. The box-type ways offer rigid, sturdy mechanicals. The flat linear-ways move easily, but naturally compromise some rigidity.

8.4   Backlash

A modern machine tool is built to exacting specifications. Surprisingly, a normal part of this specification is a small gap, or looseness, between the parts of the machine. The requirement for movement and lubrication dictates that all moving parts fit together with an acceptable degree of fit.

A gap found between two moving parts is described as *backlash*. When a mechanical system is moving, the gaps are closed by the pressure between the driven parts. When the system stops and reverses back the other way, any backlash between the parts is taken up before the entire system once again starts moving.

When a change in direction is detected, the NC computer automatically compensates for backlash by inserting or removing a set parameter value of backlash pulses. When a system becomes overly worn or damaged, the compensation becomes

1 An update on the status of linear motors is given by Stephen Czajkowski, Dr. Boaz Eidelberg, Dr. Gerhard Heinemann and Chris Stollberger "Linear motors: The future of high-performance machine tools," American Machinist, September 1996, 44.
2 An interesting and thorough article on this subject is given by Chris Koepfer, "Which Ways?," Modern Machine Shop, April 1994, 100-110.

too large and is not effective—mechanical repairs must be done.

A backlash parameter is available for every controlled axis. Adjusting these parameters requires finding the value of actual mechanical backlash. A backlash finding discussion is given in the Diagnostic Tools.

Accuracy held by all the machine-side components is exceeded in *resolution* by the numerical control computer. A control resolution of one-tenth is found on machines having a two-tenths overall specification for positioning. Normal values for accuracy, repeatability and mechanical backlash are spelled out in the OEM specifications.

8.5   Circle Cutting

A circular move involves the simultaneous mechanical movements of two axis ball screws, two guide-ways and the overall machine fixtures. By changing the backlash parameters to zero and cutting a circle, the true magnitude of machine backlash under tool pressure is etched into a hard-record test piece.

Inspecting the final shape of this circular metal disk highlights the mechanical and electrical condition of the machine. Any adjustments made to the machine are plainly documented in the final shape of the circular path.

8.6   Turrets and Tool Changers

Several different cutting tools are called up to cut a single part. A roughing tool hogs out the material, followed by a few drilling cycles, tapping, threading and finally, the finishing pass. The ability to quickly change between cutting tools is accomplished by means of a turret on a lathe and a tool changer on a machining center.

These electro-mechanical systems have a host of carefully aligned positioning switches and closely-timed event sequences. As they wear, slight mechanical changes will introduce timing delays that can hang up their operation. The instructions given in the machine drawings for setting these gaps are followed to the letter. One bad adjustment on a tool changer can wreak havoc for years. The machine makers occasionally distribute their production jigs for use in setting mechanical systems back to factory original in the field.

8.7 Relays and Contactors

Control and machine builders both make selections for electrical relays and contactors during their design process. When the machine hits the field, all these relays and contactors are separated into machine-side and control-side functions and handled accordingly. Machine relays actuate hydraulic solenoids, coolant pumps, hydraulic pumps and other ancillary systems installed by the machine builder. Asking a control-side engineer about these circuits usually results in a quick referral back to the machine-side experts.

8.8 Hydraulics

Powerful clamping and rotating requirements of a machine are performed using hydraulic power. Features include hydraulic motors for turning a tool carousel, or pistons for clamping a tool holder.

Hydraulic systems use non-compressible fluids pumped under high pressure to actuate a piston. The piston action clamps a chuck or rotates a turret. Initially, a powerful pump supplies the hydraulic pressure source, accurately maintained at steady pressure by a regulator. A dial pressure gauge reads the pressure. This hydraulic fluid is transported throughout the machine using high pressure braided hose, or flange-sealed metal tubing. When hydraulic solenoids are electronically switched, they send pressure down the hydraulic hose to the desired piston on the machine. Care is taken to avoid accidentally breaking or damaging these hard-to-replace hydraulic lines.

An interesting application of hydraulics is the hydraulic counter balance. The weight of a vertical axis is compensated by a long pressure piston, instead of the usual chains, pulleys and counter weights.

8.9   Tooling

Catalogs are sent to every machine shop containing endless tool shapes, tool holders, insert materials and hefty price tags. Expertise in the area of tooling is needed to solve the problems presented by metal cutting.

Cutting tools are designed to cut certain materials within a specified range of cutting feed and material speed. The correct "feeds and speeds" optimizes both tool life and cutting results. Violating the correct combination of tooling and material results in either broken tools or melted material.

Standard tool holders have quick release mechanisms for replacing the tiny cutting inserts after they become dull or chipped. Cutting generally goes to hell with dull inserts in the holder.

Selling tools is a big business, with no shortage of able salesmen. They keep track of the latest developments in tooling and can suggest the best tooling solution at the lowest cost.

8.10   Fixtures

Holding a part in the machine is accomplished using some type of clamping fixture. A manual vise may hold a blank part securely to the table of a machining center. A formed three-jaw chuck will clamp parts spinning in a lathe. As cutting begins, a secure fixture is essential to keep the part from moving around, or flying out all together.

If the fixture moves during cutting, the part comes off the machine as a piece of useless *scrap* metal. Scrapping one part out of a hundred is an unacceptable rejection rate in most applications.

Machining irregular shaped castings present a special fixture challenge often addressed with tapered self-centering pins, checking between casting lots and hardened form shaped chucks. Holding especially tight tolerances is another challenge for the fixture set-up.[3]

8.11   Coolant

The process of metal cutting builds heat between the cutting tool and the material. Machine tools cool things down by spraying a water-based fluid on both the tool and the material while cutting chips. Generically, these cutting fluids are called *coolants*, or "metal working fluids," and are usually water-based (or water soluble), but occasionally a shop will use an oil-based product.

When they are fresh, water based coolants smell good and work well. As the fluid gets older, they start growing bacteria and assorted nasties that smell rotten. Servicemen dread visiting the shops running rank coolant. After such visits, the skin can turn red and itchy and then peel off in the next few days. Additives for reducing bacteria, preventing rust, and extending coolant life are available.

With oil-based coolants, the machines certainly will not rust, and some advantage to part finish is claimed. However, to the discomfort of the operators, a fine mist of oil gets into everything, including the motors, machine cabinets and switch panels.

Concentrated coolant is usually mixed with water in proportions given by the coolant manufacturer. As the coolant is used, water is added to replace for evaporation. Small, handheld reflectometers aid in evaluating the coolant for proper water content. Active coolant evaluation systems are available.

---

3   Holding production tolerances is reviewed in an article by Peter Ackroyd, "Forget about Tolerances! It's your Cpk that Counts," Modern Machine Shop, April 1993, 56-61.

To extend the life of coolants, recycling systems are sold. These systems restore the proper chemical balance to the coolant and filter out contamination. One contaminant is the lube oil that seeps into the coolant after lubricating the screws and ways. Oil skimmers can pull machine lube off the surface of the coolant, both extending its life and cutting down on airborne oil. Worn mechanicals increase the amount of lube oil consumption and coolant contamination.

8.12   Cutting Chips

What does one do with all those metal cutting chips? They come in all different sizes from powder to ribbons and manage to work themselves into everything. The best chip removal systems are built into the machine design: coolant flushes the chips down into the bed of the machine, and a chip conveyer slowly pulls the accumulated *swarf* out of the machine into bins awaiting recycling.

Metal chips and electronics don't mix. Handling electronics in a shop environment risks stray metal chips getting lodged and causing a mysterious short. During EPROM changes and board replacements, an area is first cleared of debris and clean service manuals or cardboard is laid out.

If stray cutting chips litter the area, a stiff brush is used to clear the machine off before opening up the machine cabinets. Stray cutting chips found inside the cabinet are not touched. It's better to have the immediate problems fixed first, then refer the serious contamination problem to another day, when everything is running well and the shop has some lengthy production time to spare.

Cleaning away stray chips laying inside the cabinet invites an expensive loss of circuits getting shorted. Risks are explained to the person who has to pay the bill, before beginning. One small area is cleaned at a time with a vacuum cleaner (never with an air hose, and never, NEVER with the power still on).

Then everything is buttoned back up before powering up to see what shorts out. If everything is okay the next mess removal step is taken. Of course, when all the cabinets are kept sealed, like the factory intended, this needless headache is avoided.

Metal chips are never handled or tugged barehanded— they are sharp as a razor. Proper chip size and removal is addressed by the tooling and cutting conditions. During some types of cutting, the chips come off in long ribbons which can snag on a tool and spoil the finish.

In the front cutting area of the machine, chips get packed into everything. This is normal. Periodically the covers are removed for cleaning. Chips packed around switches and ball screws can cause pesky, intermittent problems. After cleaning, the covers are carefully sealed up to extend the time between the cleanings. Finally, the machine gets checked out to see if everything is running.

CNC QuizBox

8.2   A ball screw with a 10 mm pitch is turning at 1000rpm. Find:
   a.) The traverse rate.
   b.) The pulse frequency for a 2500 pulse feedback resolution.
   c.) The traverse distance for one motor revolution.
   d.) The traverse distance for one feedback pulse.
   e.) The scaling factor for a one micron (0.001 mm) "friendly pulse".

8.3   What are the advantages and disadvantages of linear induction motors as they apply to machine tools?

8.4.1   Sketch a circle test cut taken off a machine that has zero backlash on one axis and 0.0004 inches on the other.

8.4.2   Mark 16 points around the circumference of a machined circle that best describes its overall condition.

8.10   What is Cpk? If a fixture introduces a 5% rejection rate, what is the swing in the Cpk?

9 *Optional Features*

9.1 Introduction

The vast majority of CNC's in the field today are the heavily proprietary or "closed" systems. To add or upgrade one of these machines takes a coordinated (and expensive) effort by everyone in the service network. The movement is under way to "open things up", driven in large part by the promised flexibility of upgrading to the new "open" machines.[1]

Trying to circumvent the option policies on proprietary CNCs is anticipated by a plentiful use of technical roadblocks and detailed sales policies. Both points effectively manage the network of service in regard to securing optional features from the OEMs. Adding fancy options to a closed CNC system will only happen after the dealer, control builder and machine builder are satisfied that a proposed option makes sense, both technically and financially. The machine dealer then quotes the entire option price, passing along the OEM charges and any after-market expenses in the final firm quotation given to the customer.

Such options can include high speed cutting, running touch probes, or adding linear scales to a machine. In the case of linear position scales, the control builder would supply the scale interface electronics and calculate the application parameters. Engineers from the machine builder would decide the best under table installation for the scales. The after-market scale manufacturer would supply the actual scales and related technical specifications. The dealer attends with the final installation and demonstration.

9.2 Control Builder Options

End-users often make inquiries about control options. Their interest begins with the common *plug-in* accessories be-

---

1 See the references given in Section 3.2 and 7.9

fore moving into the newer and more advanced systems offered on the CNCs of today. Finding the control options available for a particular machine starts with a call to the salesman.
The salesman will run down the list of accessory control options and put together a nice written price and availability quotation.

At the time of quotation the option is fully explained and
a purchase price is laid on the table. To be sure the option
performs as advertised, it should be fully explained and demonstrated as a condition of the purchase. This ensures that a
newly purchased accessory gets its full use and operation at the
machine.

9.2.1   Factory Applications

Big dollar options exist in the coordinated *manufacturing
cells*. A typical cell application uses several CNC machines
linked as components in the collected assembly of a single
product. The manufacturing linkage between machines is
built using conveyers, robots, automatic fixtures and part pallets, all orchestrated by a central layout of carefully sequenced
cell instructions. Most pieces in the *linkage* need expensive
control options to run the electro-mechanical options.

Consider two quick examples. To maintain overall *process flow* in a manufacturing cell, each stand alone CNC must
share I/O data with the central cell network. This sharing of
data is accomplished using expensive hardware options for
shared network communications. Another option displays an
extra screen for programming the automatic storage and retrieval of pallets from a honeycomb of possible locations.

9.2.2   CNC Platforms

Before a machine leaves the show room, the choice of a
control builder and computer platform version is decided. The
major builders offer a few choices of standard CNC platforms.
Some shops stick with one or two brands for everything in the

shop, while others like to get the one best deal from everybody. Either way, once a specific version of a proprietary CNC platform is chosen, the option to change or modify to another version is essentially past. The market place will decide if this choice remains acceptable or if systems will start opening up.

Major machine shows offer the full line-up of control platform choices, from fully open PC compatibles with a mouse to the fully closed one package systems. Different levels of PC based controllers are now available for new machine applications as well as retrofit of the older controls.

9.2.3 Serial Communications

A remote PC computer linked by cable to the main NC computer provides a fast exchange of program, offset and parameter data, the same communication link lets a small laptop run a towering CNC machine with a *spoon-fed* part program.

To review, part program data is either 1) keyed in by hand, 2) loaded by tape, or 3) sent through the interface. The interface method attracts increasing attention. Paper punch systems exchanged the NC data on older machines using a standard parallel Facit data interface. Since then, builders have adopted the RS232 standard for serial communications. The RS422 standard is also offered for longer remote transfers, although the RS232 standard continues to satisfy the majority of users.

The RS232 standard interface includes settings for the baud rate, stop bits, data bits and handshaking. Both computers must be set up ahead of time with matching interface settings. To review, information moves at the interface *baud rate* setting—the higher the baud, the faster the exchange. *Handshaking* dictates how the two connected computers handle the lightning fast, stop-and-start bursts of computer data.

A single RS232 *port* running at 4800 or 9600 baud is standard. Control options expand this to include additional data ports, high speed ports, data buffers and more exotic handshaking protocols. High speed ports push the baud rates up in the 38,400 to 76,800 range. *Buffers* provide an extra reserve tank for fast depleting data. They fill and refill with extra data to smooth over any occasional *starvation* of transfer data. It is worthwhile to compare standard and optional RS232 features offered by competing builders.

The data interface supplied on new machines are in another league altogether, they use *Direct Computer Networking*[2] (DCN) with standard peripheral *protocols*[3] having character transfer rates in the 1.0 M (Mega*bytes* per second) range. This compares with 0.0006 M for standard RS232 communications.

9.2.4   High Speed Machining

A catch phrase these days for almost anything found to expand a machine's cutting capabilities is *high speed machining*. One man's high speed machining is considered basic machining by another. Full blown, top of the line options from the OEMs run into the tens of thousands of dollars.

High speed machining is built up in levels. The higher the level, the more expensive it gets. Start with standard G-code programs spoon fed into the machine using standard tape mode, end up at CAD/CAM generated, point-to-point incremental tool paths run at high speed by a combination of stout mechanicals, high speed spindles, performance CPU upgrades and look-ahead, *form compensation* software routines, all well fed by a dedicated direct computer network.

Doing simulations of a shop's part program finds the appropriate level of high speed options. Application engineers working for larger dealers perform these calculations for a certain control and machine type.[4] Occasionally a new high speed

2 An interesting article on "look ahead" is given by, Todd Schuett, "A Closer Look At Look-Ahead," Modern Machine Shop, March 1996, 80-89.
3 See also Section 7.9.
4 Review typical high speed cutting simulations in the exercises.

option is installed on a dealer's show room machine to actually satisfy the customer cutting requirements.

9.2.5   Plug In Options

Upgrading a machine is often accomplished using a supporting piece of equipment from the control builder. Improvements to either the control hardware, control software, or both, are specified in most option installations.

A list of possible improvements is firmly established between the control- and machine-OEMs ahead of time. When a customer or dealer requests an option off this list, each party supplies the necessary parts for building the complete option. Control upgrades usually require on-site visits to properly install and demonstrate the new feature.

Memory expansions and high speed buffer boards consist of actual pieces of control hardware available from the factory. Adding another axis of motion to a machine relies on extra cables, motors, servos and other control upgrades to operate the new mechanical installations. Adding the option for an auxiliary axis requires optional upgrades from all the players much like the linear position scale example in the introduction.

A lengthy list of improvements involve updating the control software. They include: touch probes, indexers, the deeply internal custom-macro (parametric) programming, tool-life management, and additional work-coordinate shifts. All such upgrades are closely monitored by the builders as part of their global profitability calculated for this business.

9.3   Machine Builder Options

Builders distribute a nice color sales brochure for the machine. All the features of construction are attractively reviewed with some corner space devoted to the available accessories and more advanced options.

Machines can be customized with part catchers, bar feeders and part counters. The number of tool stations may expand. Touch-probe tool setters may be added. Even high pressure, through-the-tool coolant systems are available. Other big ticket items include automatic pallet changing, touch probes with digitizing software and high-speed/high-power spindle systems.

9.3.1   Accessories

The machine-side option for quickly "touching off" the tool offsets is standard on one machine and optional on another. A current list of functions from the machine builder explains the basic, optional and not-in-your-wildest-dreams accessories. As a rule, options are applied to machining centers and turning centers differently.

Take the lathes first. To perform two-sided turning and milling operations within the confines of a single turning center, accessories for a lathe include live tooling and sub-spindles. Three-jaw chucks and collets are offered in a range of standard sizes to accommodate different applications. The bar-feeder interface, and possibly the bar-feeder itself, is offered to keep a lathe running non-stop production. A dandy part-catcher and part-counter is offered to keep track of all those finished parts that the option for a bar feeder ceaselessly generates.

To satisfy the demand for productivity and manufacturing flexibility, the makers of machining centers offer expanded tool pockets, pallet shuttles and the purchase of optional forth and fifth rotary axes. The specification for positioning scales is applied at the factory assembly line and seldom in the field— at least on the full-sized CNC machines.

A final set of interesting machine options involves the coolant. Coolant routed through the spindle cartridge and out the tool tip provides better cooling at higher cutting feed-rates. Additional coolant lines are installed to create a "curtain of coolant" for applications needing dust control or flushing away of the cutting chips.

9.3.2   Advanced OEM Options

Some builders can offer expanded optional features because their machine has a foundation appropriate for the job.
Top end machines are always better candidates for running advanced options. Without a good backbone, a high-speed, high-
accuracy, high-power options may be short-lived.

The overall success at running a customer's application
falls heavily on the machine maker. They often supervise installation of the high-speed cutting options even though it is
the control builder who supplies all the high speed interface
and CPU upgrades.

A machine builder retrofits a machine for ultra-high
speed spindles by going out to the warehouse, pulling out the
standard motor and drive, and dropping in a screaming 50,000
rpm unit. Exchange procedures come from the factory's past
experience with such units, and occasionally, a factory engineer is flown in to supervise the operation.

An impressive array of cutting-edge machine tool options
result when the actions of super-accurate touch probes are
combined with creative software programming. Automatic
systems can set tool offsets, align fixtures, and even create new
G-code programs from digitized samples. The software and
touch probes are supplied by a growing cadre of after-market
consultants and manufacturers.

9.4   After-Market Options

Reading the trade magazines for machine tools demonstrates the scope and variety of after-market vendors. Attracted to machine shows are booth after booth of novel new
computer products, inspection systems and mechanical accessories which are geared to some niche in the marketplace. The
ranks of the after-market CNC consultants are growing and becoming ever more specialized in their subjects.

9.4.1   Computer Software

New computer software packages are revolutionizing the way of business. The impact of integrated CAD/CAM software has changed the face of CNC machining. The days of skilled manipulation of arcane computer codes at the keyboards of clearly divergent controls are over. The age of after-market computer software has arrived for machine shops.

A visit from a software representative can best demonstrate the capabilities of the latest version of software written for the shop computer. New production software can automate CNC toolpath code, tool lists, geometry and dimensioning of new or existing jobs.

The generated NC code from one package adjusts the feeds while interpolating sharp arcs—a similar "form compensation" feature is offered by the control builders. Shopping based on this point alone is worthwhile.

Running the machine remotely or just exchanging data into NC memory relies on the shop's serial RS232 network of cabling and computer software. Simple systems use a cable, an inexpensive desktop computer and a box of communication software to effectively run the machine in tape mode, with software handshaking, at 4,800 baud (and above).

Multiple machine networks have cables running to a central rotary switch box for selecting the machine needing a download. Fancy electronic networks will install live nodes at every machine to access a local area network right from the machine tool. Many after-market companies offer to sell, install and support many different versions of these same systems.

9.4.2   Accessories

Catalogs overflow with new gadgets and inventions for the CNC business. A simple rotating fixture placed on the table of machining centers is available in electronic, hydraulic, mechanical, pneumatic and even coolant driven versions.

Trade magazines are loaded with great ideas for the shop floor. A quick review of one popular shop magazine gave a lengthy list of accessories[5] sold for machine tools. The software, tooling, fluids, optical comparators, air gauges, micrometers, gauge sets, and so forth all serve a vital purpose around the metal cutting business.

9.4.3   Consultants and Services

One consultant was hired for three months to analyze a manufacturing cell having five machines. The assignment was to speed up the cycle time and lower the rejection rate from five percent to below one percent. After three months, the recommendation for slender boring bars and solid chucks solved the problems. Another consultant specializes only in writing parametric programs for touch probes, he works full-time creating special application macros at the customer's work site.

In addition to consultants, more companies are supplying mechanical and electrical replacement parts for machine tools. Especially interesting is the quick turn-around PC-Board repair companies that repair factory original equipment at roughly half of the normal factory price. Considering that an old spindle motor can fetch $10,000, the old items are closely guarded. Companies that retrofit, rebuild or remanufacture old machines can make out on the leftover parts.

"After-market" networks of service are on the increase. Independent training classes are offered in hotel hospitality rooms, OEM showrooms, or tailored for on-site presentations. Classes are offered in CNC programming, shop management and slick optional features.

---

5   Looking at the subject of coolant products showed advertisements for coolant: dispensers, skimmers, recycling, calibration, mist spray, management and filters.

CNC QuizBox

9.2.3 How long would it take to transfer a one million character program file:
a.) At 4,800 baud.
b.) At 38.4 Kbaud.
c.) At 8.0 megabits-per-second
d.) At 1.0 megabyte-per-second

9.2.4 Specify the least expensive level of "high speed cutting" that will satisfy the customer's two cutting requirements:
a.) A segment of X.01 Y.02 F100, in a 30Kbyte part program.
b.) A segment of X.01 Y.02 F200, in a 1Mbyte absolute part program.
c.) A segment of X.001 Y.002 F500, in a 60Mbyte CAD/CAM generated, incremental part program.

Select from the following options:

Option	Name	Cost	Specifications
1	Standard NC	Basic	Block to Block processing=6ms Feed≤200in/min, Baud≤9,600
2	High Speed CPU	$10,000	2ms, 600in/min and 9,600 baud
3	DNC Level 1	$2,000	Baud≤38.4K
4	DCN Level 2	$4,000	Baud≤10M with Buffer=16Kbyte

2nd Tool

Diagnostics

Introduction

Just swapping parts is one consideration for fixing a machine. However, to repair means to dig into the problem, diagnose the cause and formulate the treatment. Like a doctor, the serviceman makes a house call to perform some indicated tests on the ailing patient. After the easy guesses are tried, it's time for the *Diagnostic* skills.

When a person becomes sick they use built-in senses to explain a headache or stomachache to the doctor. These are *internal diagnostics* supplied by the patient. A machine tool also has internal diagnostics that are built in at the factory. By calling up information from the operator's panel, or other visual displays, a serviceman can "talk to the machine" and identify the nature of the problem.

When a patient describes mysterious, unknown symptoms, the doctor requests a battery of *external diagnostic* tests be done. A blood test, treadmill, heart monitor, or exploratory surgery is conducted to isolate the trouble. In order to have enough information to prescribe an effective treatment, even more testing may be indicated by the initial results.

The external diagnostics of a machine tool are the test systems connected to the machine to monitor its operation. Results of external test equipment and test programs supply the healing data for the serviceman. As data comes in, additional tests are suggested to zero-in on the trouble.

A doctor's knowledge is learned in medical school through books, demonstrations and lectures. Diagnostic tools are appreciated in treating real patients during their internship. New and better diagnostics are learned as their experience expands. A successful "machine" doctor uses data from both the internal and external diagnostics to help solve a machine problem.

10 Test Equipment

10.1   Introduction

A final repair, documented with solid test equipment data, provides a confident result for the customer and other involved engineers. Choosing the correct "external diagnostic" test method is based on the symptoms observed and the testing skills available. Of course, absolute safety and damage prevention is the highest priority.

Before choosing a test procedure, a few questions are asked: What is the purpose of the test? What are the key variables or data actions of the test? What are the expected results? How will the data be recorded once the testing begins? Should someone man the main power cut-off switch at all times during the test?

These questions deserve consideration before a test is conducted. Tests must anticipate any possible problem. The equipment is installed from a carefully written procedure while the machine is in a power-off condition. All test leads and power cords are secured and insulated before the actual power-on testing begins. The layout is checked and double-checked for each specific test before power is ever applied.

10.2   Voltmeter

The trusty voltmeter, in the right hands, is a service machine. A new meter has instruction manuals outlining its proper use and interpretation, but experience serves as the best teacher for capitalizing on the meter's true capabilities. Voltage, current, resistance and their dynamic data trends provide the insight needed to solve complex electrical problems.

Two basic types of meters are available—the analog meter having the moving needle display and the powerful digital multimeters with LCD (Liquid Crystal Diode) readouts. If time is taken to learn the special functions of the digital meters, such as "auto-ranging" and "min/max data capture," it greatly expands their usefulness. Accessories for measuring temperature, vibration and rpm are widely available.

Data displayed by the voltmeter is carefully recorded, making sure to include any electrical units involved and any prefix or decimal points. Missing one feature usually renders the data meaningless. Understanding the concepts of voltage and resistance helps the user capture more meaningful data from a meter. Understanding how a meter detects and displays data is also helpful.

Experienced service engineers will gingerly connect test leads one by one, with the power off. With a firm balanced footing, connections are carefully made with one hand, while the other is kept free of the test circuit. The free hand stabilizes the body position and thwarts one path for deadly current. Any worn or damaged test leads are immediately replaced with a nice fresh set. Later the bad leads are either repaired or discarded. The hidden dangers always remain, even when the power is safely disconnected.

10.3   Test Clips

A good selection of test clips is a valuable part of any tool kit. Alligator clips and test probes in several sizes and configurations are available. Sets of full-size and miniaturized alligator clips, connected to long jumper pigtails will handle the basic test set-ups. Another set of retracting hook-connector probes are handy for getting into tight places, behind printed circuit boards and the like. These tiny hook probes clip onto a leg of an integrated circuit, where the alligator-style clips would never fit.

Flexible plastic insulator boots found on test clips can open a gap that will short out a test circuit. To prevent this from happening, test connections are insulated and secured by generous usage of high-quality electrical tape, long before a test circuit is charged up. If any components are in danger of shorting together, another test point or a better-suited probe is found. Clips and probes are kept together, neatly stored in a little bag, safely tucked in the back of the tool box.

The supply of test clips, like voltmeter probes, are kept fresh. If a clip is getting frayed, the wire is immediately clipped off and stashed for future repair or replacement. It's ridiculous to goof around with beat-up, two-dollar test clips when thousands of dollars of needless damage could result.

10.4   Oscilloscope

Setting up a nice picture on an O'scope gives quite a show to the customer. When properly explained and interpreted, the *oscilloscope* signal data leads to clear-cut service decisions.

In general, the scope is used to check continuous, periodic signals with a trigger, or those signals slow enough to be viewed in real time. The scope displays time and voltage on a small, graduated screen. The time scale is read along the horizontal axis, the voltage levels are read on the vertical axis. The captured signals rise and fall while traveling along in time.

To calculate a signal's value, the screen is overlaid with criss-crossing, graduated lines that form a grid. Each line marks a *division*. If the details of a display become too small to see and measure, it's simply expanded in time or voltage by turning the scaling (per division) dials.

The signal type and trigger must be correctly applied to get a steady, meaningful display—these concepts require practice. A repetitive signal, like the simple AC signal from a wall socket, is easily captured ("triggered") on an O'scope, but a single shot event only flashes the screen without leaving a lasting trace. Suitable selection of test equipment depends on the type of testing at hand.

Most scopes are large units intended for use on a test bench. Dragging these units around on service calls is tough and leads to their early demise. A better choice for field service is the compact, portable units carried with a shoulder strap.

In practice, the voltmeter gives a number, and the oscillo-scope gives a picture. Confirming voltmeter data with an O'scope raises confidence in the repair data and might result in a new conclusion.

10.5   Recorder

Not that long ago, an O'scope in the field was considered a luxury, but today they are commonplace in field service. The next tool gaining such popularity is the *data recorder*. With the advent of high-speed, computer-driven, electro-mechani-cal systems, the standards for a repair have been raised.

A data recorder is like a camera—it takes a snapshot at precisely the right moment of any signal accessible to its test leads. They are used by the OEMs during the original design and testing of a new machine to document the key signals of a normal machine. By comparing the signals of a troubled ma-chine with the known factory standard, a highly complex prob-lem is quickly identified and solved. Such blind data comparisons increase the use of data recorders in the field.

Economical two-channel data recorders are now available with plenty of power for use in the field. Captured data is dis-played on an LCD screen and then stored in resident memory. Once data is held in memory, a transfer to a lap top computer or paper printer is possible. Many times, data recorder results faxed in for factory analysis closes a service (right away). Cru-cial data is right there in black-and-white! Every expert in the network will appreciate its meaning.

Much more expensive units used by the OEMs have more analog "signal" and digital "data" channels and a built-in pa-per printer. Trigger controls are also more easily modified for level, delay and channel. However, their size and expense dis-courages field use except in the most critical of applications.

Established test equipment manufacturers offer classes on using all types of test equipment including data recorders. Like an oscilloscope, the data recorder displays a time-varying signal. The O'scope displays periodic signals beautifully once a trigger is properly selected, but it falters at one-shot events. For one-shot events data recorders really come into their own. A brief example may help understand a recorder's usefulness.

Suppose a servo alarm occurs every once in a while and the motor torque needs to be captured right before the alarm occurs. A data recorder is easily capable of this performance. In this case, the data recorder triggers off the alarm occurrence and displays the stored shape of the torque data leading up to alarm. The central question is how to trigger the data recorder with the servo alarm. Finding the snapshot trigger always presents the biggest challenge.

To begin, the list of signals that immediately change when the alarm happens is considered. In this case, the servos are immediately shut off in response to the alarm, these shut off signals are a perfect trigger for watching all the other signals on the machine.

The trigger is the key. With a good trigger found, the data channels are moved from signal to signal creating a portfolio of actions leading up to the servo alarm. Is this problem caused by excessive load, current, or torque? Is it excessive motor speed, or did the DC power supply drop out? Perhaps some totally unexpected signal is involved. Regardless of the cause, eventually the solution is presented clearly in the data recorder's indisputable signal snapshots.

10.6   Lap Top Computer

Everyone needs a lap top these days. A unit having a fax modem and communication software can communicate with the machine over the serial interface, conveniently sending and receiving G-code programs, testing the interface and saving and verifying the NC data.

New diagnostic test equipment also has RS232 capabilities. Captured test data is directly sent into a lap top for software analysis, or transferred to a printer for a hard copy print out. Rapid transfer of test data over a phone modem is another practical option with a good lap top computer.

Computer based training and on-line maintenance systems are accessed over the public Internet, or through country-club access to the protected servers. An interactive exchange of ideas with several machine and control experts, all finished off with a parts shipment, is changing the face of service. Information is itself a product, rapidly being delivered to every waiting customer in the field.[1]

10.7   Specialized Equipment

Other kinds of test equipment are applied for special testing. A *meg-ohm* meter checks the high voltage insulation of a motor. A *stethoscope* can listen for worn bearings. Dummy connectors plug in to simulate a missing load or to temporarily remove a signal condition for testing. A handy break-out box can quickly check serial communication signals.

10.8   Machine Side

Machine shops already have equipment on hand to inspect their production parts. Such precise physical measurements, with micrometers and dial indicators, can check a machine's positioning accuracy. Machine calibration performed after a major mechanical repair uses everything from expensive laser positioning equipment to carefully set indicators.

Getting close readings of a machine's accuracy relies on several approaches. Sweeping a circle with a spindle-mounted dial indicator, or taking micrometer measurements of a rectangular test plate is both quick and effective. Service technique

---

1 Ask the CNC experts is a new service offered at several web and usenet sites. For usenet access connect with alt.machines.cnc or get further instructions and locate the on-line references for this book at www.cncbookshelf.com

checks a machine's level, positioning, repeatability and squareness. More specialized investigations employ the DBBs.[2]

The trusty dial indicators come in different increments: one thousandths, half-thousandths and tenths are common for mechanical dial indicators. Portable mounts hold these gauges firmly in place using powerful magnets. Gauges are expensive and easily damaged, especially while mounted inside the movement plane of a machine.

When an overall machining process has excessive total error, the final parts are mysteriously *out of print.*[3] It's tough to nail down such a problem because the total error is an accumulation of many smaller errors introduced by the process. Key questions are: How accurately is the part clamped? Is the machine cold? How accurate is the inspection? Is the machine accuracy good? Is control accuracy good? Is application and tooling good? The service people have a full plate.

Looking at a final "out of print" part is the first step. Going back and checking each link in the cutting process suggests the combination of minor events leading to the problem holding part dimensions.

10.9   Pen And Paper

All the fancy and expensive test equipment in the world is no substitute for a carefully written set of observations. Before considering an approach, a written game plan is necessary. When the game plan includes using external diagnostic test equipment, a sketch of the intended check points and blank "fill in the data" tables are first proposed and prepared.

2   An article that mentions the ASME (American Society of Manufacturing Engineers) machine inspection standards and reviews checking a machine with, among other things, a DBB is given by, Rick Glos, "Do-It-Yourself Machine Inspections," Modern Machine Shop, April 1996, 56-64.
3   Further machine inspection and Double Ball Bar references are given in Sections 1.13 and 15.3.

The power of the pen and paper cannot be overemphasized. Visualizing the goal of a test becomes clearer during the process of writing it down. A clear test description, written down beforehand, stakes out the area covered by the test and helps avoid silly and often costly mistakes.

If the notes of an ongoing service are kept clear and concise, repair can move forward. Sometimes stepping back and rethinking the situation gives the necessary time for new ideas to develop. Good, readable repair notes can also be passed on to others more qualified—an important step in making a successful NC repair.

10.10   Schematics

Deep secrets of electronic control circuits are exposed while following the electron road map of a schematic. The written electronic symbols and wire lines hold such a fascination; however, in the hands of an inexperienced observer, they can lead to nothing but trouble.

The different levels prepared for NC machines are: elementary schematics, internal schematics and machine-side schematics. All contribute to describing the CNC system, as a whole.

Machine-side schematics show major wiring for the switches and solenoids used on the machine. The model type for each major machine component is listed, and each corresponding wire and terminal strip has its tag referenced in these particular schematics. Replacing a solenoid or switch is greatly simplified using the *machine-side* schematics.

Elementary schematics show the relationship between printed circuit boards inside the control and also build a link out to the circuits in the machine schematics. The control modules—the power supplies, computer boards, servo-units, motors and operator panels—are all linked together in the *elementary* schematics.

Internal schematics show the deep board level wiring and components. Jumping in with the *internal* schematics before a problem is fully understood constitutes grand folly. A copy of the internals is normally kept at the OEM. Some makers allow for distribution of the PC-Board internals and others will not. If tiny board-level components are not available from the service department, they generally will refuse a request for internal schematics. Internals prove useful in only a fraction of field service repairs.

Every new machine is delivered with elementary and machine-side schematics, and they are valuable external diagnostic tools. It is important that the original copies stay with the machine. Hunting for schematics while the machine is down distracts the service process, and if the schematics are lost, it severely frustrates the approach.

10.11 Connecting Signals

Gaining access to an electronic signal sitting on a big, protruding screw terminal is easy with a sharp-toothed alligator clip, however, getting at the same signal running inside a bundled connector of 50 fragile sealed wires is almost impossible. Difficult to open and tricky to probe, a tiny signal wire is a test that challenges the skill of a repairman.

Placing the test equipment starts by finding the signal of interest in the elementary schematics or servo application manual. These manuals give the connector names and pin numbers printed out on the machine.

The connector name is found printed on a connector or to the board in which it's plugged. The pin numbers within the connector are found using a bright flashlight and reading the raised plastic number molded into the top or bottom of a connector. When no numbers are found on the connector, tracing a few wires to known circuits will establish the correct counting order of the connector.

Monitoring the wrong signal on a connector gives amazing, albeit worthless, results on the test equipment. Everyone with service experience has a preferred method for installing test equipment without destroying any wires or connections.

Before powering up the system, an experienced service engineer will double-check everything to be sure no disturbance was introduced by the application of test leads, and all the hidden dangers are realized. Necessary data is captured, and when the test is finished, the machine is switched off. Test equipment is then carefully removed and any disturbances to the system are taken care of with electrical tape and miniature tie-wraps.

10.12   Safe Connections

Experienced service engineers understand that some preparations are in order before a test connection is set up on a CNC machine. A proposed list of diagnostic equipment is identified by type, make and model. Additional notations are made to describe the proper signal polarity, data check-pin locations and labels. The machine schematics show the signals and locate the exact connectors and wires for those signals. Documentation is drafted that describes how each diagnostic test channel is tied to the CNC signals under testing, with a few explanatory drawings thrown in for good measure.

Carefully, the requirements placed in the drawings are transferred to the machine under test. When all the test probes are connected, insulated and secured, the area is cleared in anticipation of returning live electrical power to the machine. All test equipment is clearly readable from a safe distance, and no moving of the test set-up is considered once the power comes on.

CNC QuizBox

10.5 Using a lap top computer having a 4 Gigabytes of memory, 32 Megs of Ram and a 56K modem, all running communication software at 300Mhz.
  a.) How long would it take to transfer a one million character program file over the modem?
  b.) How could this computer gain access to the internet?
  c.) Name at least six service tasks accomplished using a lap top computer.

10.7 List seven items that contribute to the total error of a machined part.

11 Electric Power

11.1   Main Power

To reliably run machines on electricity, the OEMs issue clear specifications for the electrical power. The *main* power is a three-phase (3φ), balanced, alternating source sent to the machine in either a delta or wye configuration. A wye source has three live legs and a central neutral conductor, while delta has three live legs and a floating neutral. In both cases the live legs are commonly assigned the letters R, S and T.

The amplitude between each leg is 200/220 Vac rms with a voltage swing of plus ten or minus fifteen percent commonly allowed. This puts the ideal amplitude in the range of 242 and 170, on average, 206 Vac per phase. Such exact numbers will vary between OEMs.

An electrician visits a new machine installation to check that the main power meets with the OEM standard. With the power off, voltmeter probes are connected and the meter is taped down in an observable location. While standing away, the power is clicked on. Data is displayed and recorded, and the breaker to the machine is switched off. The electrician safely obtains main power readings with this fail-safe, power-off technique.

These three-phase voltages describe the amplitude and balance of the machine's main power. Balance and amplitude are both important while talking about three-phase power. Lengthy treatment and discussion of the transmission and distribution of three-phase power is handled in the field of Electrical Engineering.

To observe the shape of the alternating voltage on the main power an electrician must use specialized scope leads and safety isolation circuits. The peak-to-peak level shows up higher than a meter's Vac rms (Root Mean Square) readings. An electrician understands this difference, along with many other technical topics, like the deadly hidden danger associated with using a scope's test channels on 3φ AC circuits. As a rule, standard scopes are not used on any high voltage AC or DC circuits. Fixing a problem includes locating the right person for each phase of the job.

Monitoring the main power shows the effects to the voltage when the machine is put under load. Simultaneously moving several axis motors increases the current demands of the machine. If the demand is not met, the amplitude of the voltage drops. Some fluctuation is normal, but excessive drops outside the swing range suggest a sub-standard service to the machine. A licensed electrician can also evaluate the proper size for wire connections and wire gauge leading up to the machine.

Electrical parts will fail repeatedly if the main power is bad. Out-of-specification power problems, like drop outs, damage NC machines. A lightning storm that suddenly knocks out the main power instantly leaves spinning motors charged without computer logic for their control. Expensive damage follows the path taken by these abandoned motor currents. With powerful thunderstorm cells in the area, a shut down of the CNCs while waiting for the main storm front to pass avoids unexpected service calls.

If one leg of the main power has a loose connection, the short-lived arcing causes *drop out* and results in reoccurring, expensive service problems. Evidence of arcing includes burn marks and melted metal connections. When the arcing is active, a glowing connection is actually observed.

Three-phase power monitors are routinely used by the power companies to provide a chart record of the supplied line power over several days. Many times a dealer or visiting OEM will check the main power and answer any questions about the acceptable levels given in the machine's specification.

11.2   A Hundred Volts-AC

The voltage source common for household appliances is around 100 Vac; this type of single phase (1φ) power also runs many of the relays, solenoids and cooling fans on a CNC machine. To arrive at this common single phase 100 Vac level one phase of the main power (usually the R to S phase) and a large step-down transformer is used. Monitoring the AC100 shows a nice scaled down replica of the 220 Vac main.

If the AC100 should ever drop, the effect is dramatic. Imagine all the relays suddenly de-energizing. When the coil voltage of the relays drop, the magnetic field drops, which allows the contacts to release, losing the switched circuit.

Machine-side schematics give the best check points. Normally, a few pages of schematics are devoted to the AC100 circuit. The most accessible terminal screws for connecting test equipment are found by asking for advice and schematics from the OEMs.

11.3   Power Supply-DC

Control makers receive calls for replacement power supplies all the time. Power supplies are asked to do a lot on machine tools. Voltages from these units are used for computer logic and must be closely regulated for uninterrupted logic control. Voltages also travel out to the I/O circuits and operator panel on the machine.

As DC power travels away from the power supply, the level drops slightly due to the intervening wire resistance. The voltage level is checked at the end of the line with a voltmeter set for DC to get the levels, and AC for the ripples. The elementary schematics list the possible checkpoints.

Level adjustments are available inside the power supply for certain outputs. These are **never** adjusted until all computer data is recorded and backed up.

A small, small turn results in large, large swings in the DC level. Too large an increase causes the signal to burn out the PC-Boards, and too large a decrease causes the NC computer data to become scrambled. Adjustments are only considered by a service engineer when something is broken or has been replaced. The acceptable range of DC outputs are specified by the OEMs in the machine maintenance manuals.

Problems with the power supply will trigger an alarm or blow a fuse. Elementary schematics show the involved fuses and alarm contacts. A voltmeter placed across these alarm contacts indicates the current alarm condition. (When contacts are closed the voltage drop is essentially zero. When alarm contacts open, a signal voltage develops equal to the voltage level of the I/O switch monitor).

The elementary and machine-side schematics laboriously trace down the source of power supply problems. The OEMs hear about and rattle off the answer to these questions daily.

11.4   Servo Loop

Two types of power run the servos. The big stuff is used for motor current, and the small stuff is used for the logic control circuits. Again, the elementary schematics explain the power connections into the servos.

Small power for the servo controls is supplied first—this logic power gets everything under control before the big power is sent in. The currents involved with the motors are bigger.

11.5   Spindle Loop

Spindle drives usually have their own internal power supplies that create DC levels for the control logic circuits. Once again, control circuits are first powered to put everything under control before large motor currents are constructed from the main power input.

A spindle drive with bus capacitors is dangerous long after the power is removed. Never work around a drive, when spindle systems have a problem bring in someone with thorough training from the OEM.

11.6   Control (1st Power)

When all the computer and logic circuits of an NC machine are running, the control screens light up and all the displays and PC-Boards have active logic control. The power supply and drive units are also powered to their first level.

A magnetic switch contactor starts the distribution of power to all these control components. The elementary schematics will show the first power contactor (NC-M). When it is energized, the systems can check for troubles, and if necessary, activate protective alarm functions. The main circuit for all the servos is still disabled, so the machine cannot move around until the next level in the power on sequence, called second or servo power, is achieved.

11.7   Servo (2nd Power)

Servo power is the second and final level in the power-up of the machine. Any problems at first power will stop the machine from getting to second power. Servo power activates the big currents of the machine and releases any axis motor brakes on the machine. Second power circuits can include axis motor currents, spindle motor currents along with machine-side pumps and solenoids; those circuits the machine maker specified during the initial design of the machine.

A large three-phase switch contactor provides the central source for servo power. This heavy duty switch is called the servo contactor (SV-M) and when it is energized, the machine is ready to go; everything is "up and running." Occasionally, the main current contacts become welded shut from internal arcing. A bad set of welded contacts will never open, causing a chance for more damage at the next power-up of the machine.

Inside the servo contactor are delicate alarm contacts which give the switching status. These same contacts provide a good trigger for test equipment when some alarm triggers a "servo off" condition. When the emergency stop button is pressed, SV-M usually (not always) trips off. Large currents are *cutoff*, and the machine drops back down to first power.

11.8   Power On Sequence

A collection of steps designed for bringing up the machine first thing in the morning is called the power-on sequence. By applying power in a specific order, numerous safety functions are achieved. Sending in main power at the side of the machine is the first step in the sequence followed by control power and then servo power.

This sequence is never disturbed. The OEMs can investigate at which point the sequence stops and locate the reason over the phone. Forcing power into the machine can cause quite unbelievable damage. Manually energizing contactors is also a very bad idea. Fixing a problem includes locating the right person for each phase of the job.

A power-on sequence problem is presented to the OEMs daily, and with this experience, they handle such problems quickly over the phone. The elementary schematics will also explain exactly how the sequence works.

CNC QuizBox

11.2 An oscilloscope displays a peak-to-peak AC voltage of 308 Volts using the screen graduations. What would a hand held digital voltmeter read?

11.3 A hand held digital voltmeter read 100Vac rms across a 60-cycle test circuit. Sketch the picture an oscilloscope would display with the time and voltage divisions correctly labeled.

11.8 Draw an elementary circuit containing a series connection of:
a.) A 3ϕ main breaker having both load and ground fault protection.
b.) A 3ϕ, 220Vac input circuit converted into a 1ϕ, 100Vac circuit.
c.) A 1ϕ, 100Vac input circuit converted into a +5V, ±12, ±15 and +24Vdc output circuit. Add fuses to both the inputs and outputs and add a circuit for detecting blown fuses.

12   *Servo Loop*

The External Signals 12.1

Older
Systems
DC Servo Loops 12.2
    12.2.1 Signals
    12.2.2 Alarm Trigger
    12.2.3 Speed Command vs. Speed Feedback
    12.2.4 Signal Analysis
    12.2.5 Motor-DC
    12.2.6 Loop Feedback
    12.2.7 Swapping Parts

Newer
Systems
AC Servo Loops 12.3
    12.3.1 Signals
    12.3.2 Alarm Trigger
    12.3.3 Speed vs. Torque
    12.3.4 Signal Analysis
    12.3.5 Motor-AC
    12.3.6 Loop Feedback
    12.3.7 Swapping Parts

The Internal Signals 12.4

Older
Systems
DC Servo Loops 12.5
    12.5.1 Computer Numerical Control
    12.5.2 DC Servo Unit

Newer
Systems
AC Servo Loops 12.6
    12.6.1 Computer Numerical Control
    12.6.2 AC Servo Unit

12.1   The External Signals

Checking out the servo loops of a CNC machine is done with the help of "diagnostics." The most immediate and important diagnostic is the parts coming off the machine. When the parts have good finish and proper size, the servos are good. In the course of investigating a servo loop, two formal diagnostics are available: those internal to the machine and those brought in from the outside.

Internal diagnostics are found at the operators panel, nothing is plugged in or connected; just write down the results right off the display. Examples include looking up a servo alarm, recording the axis following error and monitoring an internal NC signal address. These "internal" servo diagnostics begin in Section 12.4 and continue in Chapter 14.

Outside loop diagnostics, or *external* diagnostics, include checking the machined parts. Inspection is a great external diagnostic—collecting inspection data from calipers, optical comparators, coordinate measuring machines (CMMs), or any other measuring equipment outside the machine. The best example, however, is the electronic test equipment brought in during a service call: voltmeters, recorders, lap tops and oscilloscopes.

To operate external test equipment requires specialized knowledge of the machine in question. Capturing useful results from diagnostic equipment is a professional skill. People skilled in these diagnostics are the inspections people, computer consultants, service engineers and active operators. They all use their knowledge and skill (diagnostics) to find and solve machining problems.

12.2   DC Servo Loops

Application of analog circuits in old DC servo loops permits direct observation of the control signals in the loop. Schematics and test equipment can directly document the loop commands and motion feedback. This data can then suggest loop remedies.

Factory loop documentation for servo-units and DC motors improved in later versions. Servo maintenance manuals include vast numbers of potentiometer adjustments and procedures for using the many board-mounted check-pins.

The references needed to verify DC servo loop dynamics are the basic motor and drive specifications combined with original NC parameter lists and elementary schematics.

12.2.1 Signals

To accommodate mass market applications, a standard system for setting servo loops is designed into the machines. The factories used this system to quickly set up or tune in different machine applications. Components in one machine's servo-loop will also run other similar applications. Motors, servos, feedback units and computer cards were all designed for as many applications as possible.

Checking the servo loop action uses the same servo check-pins and loop procedures the factory used to set up the loops. Checking the control signals within the servo loop (during test operation) documents the machine. For the old DC servo loops, the D/A-command and TG motor feedback supplies the most useful data recorder printouts. Analysis of loop following error and machine velocities follows up the investigation.

Every servo has an application manual to describe the function of each terminal, potentiometer, check-pin and power rating. Particularly helpful are the factory adjustment procedures linking board level potentiometers with monitor check-pins. Application manuals speed up signal checking by listing the best check-pins and potentiometers.

A check-pin is a small gold pin protruding from a logic control board. Each pin is identified by a letter or number like

"CH7" or "0V." Due to its miniature size and cramped spacing, the power is turned off before applying any miniature test clips. Shorting a pin to an adjacent pin or component will instantly kill the unit and ruin the whole job.

12.2.2 Alarm Trigger

Older Systems

Tough problems demand hard data to convince the customer that a problem is really fixed. This data is difficult to come by while standing behind the machine with hands in pockets after a very intermittent problem has again mysteriously just disappeared.

Imagine a hard printed record linking a problem and a solution directly to a repair action. The initial data shows the problem, and the final data reflects the finished repair. No more return visits, wasted parts, or machine down time. Before and after recorder data removes any doubt that a repair is permanent, but will require extra time and effort to set up.

Only as long as another alarm is expected is such a test set up feasible. The idea is to trigger a data recorder by the alarm occurrence. Using three channels, two are connected to the pertinent signals involved with the problem, and the third channel triggers the event. The next time the alarm is generated, the data recorder is triggered, and two involved signals leading up to the alarm itself are captured for analysis.

The alarm trigger and choice of signals to capture is the most interesting part. The trigger needs a switch contact or signal level that immediately changes when the alarm occurs. Deciding which two signals to monitor comes with the problem, and from experience. First capture the data and print out the results, then adjust, modify, or replace the defective components and wait for another trigger. When the alarm is gone and the system fixed, another problem is completely documented to never return.

The toughest can't-be-fixed-because-I-can't-see-it problems benefit from this approach. Extra time and effort is justified in special situations, like collecting and justifying the service charges to a long standing customer, or carefully documenting a new problem for others in the network, before it shows up on another machine.

12.2.3 Speed Command vs. Speed Feedback (*Older Systems*)

A powerful diagnostic for analyzing the entire servo loop on both new and old machines is a plot of the motor speed and torque while the machine is running. Subtle deviations found in this data reflect precise mechanical or electrical conditions inside the servo loop.

The speed versus torque discussion is especially helpful in analyzing the high performance AC servo loops. A limited analysis of motor speed feedback and the D/A-command is usually sufficient for the trusty old DC servo loops.

Older loops receive replacement servos and computer cards over the years. Each repair introduces slight mis-adjustments to the servo loop, which eventually accumulate into pesky, intermittent servo problems.

To go back to the factory-intended operation the loop characteristics need to be adjusted. Observing the motor speed, D/A-command, and error pulse gives a quick health check-up of a DC servo loop. Adjustments suggested by this data puts the machine operation back within factory specifications. (Signal check-pins are provided for these loop diagnostics.)

With a data recorder connected, the machine is powered up, and a small test program run to capture and print the signal shapes representing the servo loop dynamics.

Instantaneous motor rpm and D/A-commands are easily calculated from the amplitudes of the captured recorder data. Continuous values of following error are compared with the expected loop values. Each loop is adjusted and matched to balance all the loops around the machine.

197

12.2.4 Signal Analysis

Amplitudes, time scales and cycle times are tweaked to give the most meaningful data recorder printouts. A complete cycle of test data is compressed into the same picture, if possible. The captured signals are placed on top of each other, or above each other, to provide the best clarity of explanation. To capture the best, most comprehensive data traces, the servo loop testing speed and the diagnostic equipment scale settings are switched around.

A permanent record of the machine's original condition is critical before making any loop adjustments. Original signal shapes for a servo loop are first documented with the recorder. Proper adjustments are then accomplished when new loop signals can better approximate the correct factory values. Final servo loop adjustments transform the original "out of specification" data to the correct factory designs.

Specifications are available from the OEM. These settings involve the combination of possible motors, machines, servo-systems, feedback systems and NC parameters.

12.2.5 Motor-DC

Diagnostics for checking DC axis motors are limited. All motors will have brush covers, caps and possibly an access inspection plate for inspecting the armature for wear and arcing damage. The exterior of the motor is always carefully cleaned before breaking any seals to avoid the ingest of metal chips or oil into the motor. The machine power is turned off and locked out before starting a motor investigation.

Motor experts look at the face of the motor brushes to gain insight into its running condition.[1] If evidence of arcing, rocking, or commutator grooving is apparent, the motor needs

---

1 See the reference given in Section 5.6.3

service from a qualified motor expert. These old motors will usually run a while at the slower speeds, before they finally quit altogether.

The OEMs offer differing opinions for inspecting and maintaining motors based on their designs and accumulated field experience. To be safe, the original manufacturer is consulted for all motor maintenance procedures, schedules and replacement parts.

Comparing the *megger* (mega-ohm meter) readings between a good motor and a suspect motor indicates the condition of the motor. Megger readings improve after cleaning out the excess carbon dust from the motor.

A complete motor replacement is dangerous. When someone unhooks a motor mechanically or electrically, the weight, or stored energy, normally held by the motor action can suddenly come crashing down. Qualified experts block up the machine axis to hold this weight and avoid one cause of damaging drops and crashes. The original machine builder is called in whenever a motor service involves replacing a motor. The job is too heavy and dangerous for most technicians.

12.2.6 Loop Feedback   ⟨ *Older Systems* ⟩

Feedback units supply the position and speed information needed for overall loop control. Every unit provides this motion information. Of course, some fundamental differences of style exist between machine manufacturers.

Old style feedback units send a *position* pulse train and a DC proportional *speed* signal. Other signal schemes exist, like linear induction scales and motion resolvers, with a similar purpose but completely different operation. Elementary schematics show the differences (the old style is assumed here for all the upcoming discussions).

Both position and speed signals are observable using an O'scope. The speed signal check-pins are located at the servo. Position signals are picked up at the section responsible for counting the incoming feedback pulses; usually some check-pins or signal markings are found printed on the main computer cards responsible for servo loop control.

12.2.7 Swapping Parts

When the machine is acting up and the same axis is always involved, moving a suspected unit from one loop to a different loop offers a quick diagnosis. However, risks inevitably occur from such actions. People end up spending a lot of extra money cleaning up the mess created by someone recklessly swapping parts around. Many mistakes are possible, even with the seasoned service people.

The factory has the best advice for the exact version of servo loop involved. They can best explain the specific risks before choosing a plan of action. They can also determine if compatible units are available within the machine.

Common servo swapping disasters include connecting good amplifier units to shorted motors or moving just the front PC-Board on units without such a feature. Some units allow this and some don't—only the factory can say for sure. The worst is when everything gets swapped all over the place, mixed up and nobody remembers what went where.

Simple maxims for moving servos is keep the original settings in each loop and record the serial numbers and placement of all the original units. Incompatible servos are never exchanged (a mill may have the same unit on another axis, but the lathes, because of their mechanical construction, generally do not). Power is applied just long enough to see if the problem moves with the unit. Once this is determined, the machine is put back to original and a replacement secured. No sense running a machine around with the servos out of place.

A misadjustment or miswiring only results from careless procedures and a lack of attention to the original system. A pile of disassembled parts found in a box behind the machine automatically adds another digit to the final repair bill. After a fruitless swap festival, a deeper and more detailed investigation of the many signals of motion and status passing through the servo loop is needed. The servo loop was originally adjusted and documented by the factory before the machine tool hit the market, they are best able to put it back to original.

12.3   AC Servo Loops

Early AC servo loops use many of the same external diagnostics applicable to the older DC servo loops. Lately, servo improvements have transferred these previously external diagnostics into the control.

Now live displays of servo-loop currents, loads and speeds allow for quick loop adjustments without the need for bulky external test equipment. Making quick selections for different motors, feedback units and servos is available in software. Expanded following error displays capture data peaks and trends.

12.3.1 Signals

Application of motion control is big business around the world. Successful designs seldom require any field adjustment to the servo loop settings. All the gains, speeds, distances and accelerations are properly calculated at the factory.

Running the motor up to maximum speed, followed by a rapid stop and reverse to maximum puts a loop through its paces. Capturing the levels of motor speed and torque during this running cycle is a powerful design checking diagnostic. Informed analysis of this data reveals the original design setting selections.

The "big time" factories calculate settings for the servo loops after considering many complex design issues, and in the end, implemented the best solution into the machine. Be very wary of an "expert" who feels these factory settings can be somehow be improved.

A picture of the speed and torque during a test program is documented by a data recorder at the factory. The signal check-pins are given in the standard servo manual for AC drives. Some of the newer drives encode such loop signals to foil monitoring without a special signal break-out box. After the factory loop monitor is plugged in, all the normal check-pins for speed, torque and pulse feedback become available. Thankfully, most systems grant direct access to the motor speed and torque signals.

On the mixed (analog/digital) servo technology, adjustments to the loop uses familiar potentiometers. On newer, mostly digital servo loops, countless parameter addresses listed in the assigned parameter lists are adjusted. As a rule, the digital loops are still running with an unknown problem when parameter changes appear helpful. The machine worked when it was new at the factory, so once the real problems are found the original settings will work once again.

12.3.2 Alarm Trigger

The incredible performance of the servo systems built into the new machines naturally invites running the fastest, hardest and longest machining situations. Occasionally a machining operation will unwittingly cross the line and exceed the basic running specifications of the machine. The problem doesn't show up in the loop current at the instant the "overload" alarm occurs—it only becomes evident in a power analysis of the overall repetitive machining cycle. Documenting the continuous rating of the servo loop is an arduous process.

Data amplitudes and time durations lead to the continuous loop power levels. Using the "area under the curve" approach, continuous power is derived from the captured running cycle data. A data recorder captures the amplitude and time data over one machining cycle, and if an alarm is involved, a third channel can conveniently trigger the data events.

12.3.3 Speed vs. Torque

Motor speed and torque are helpful diagnostics for analyzing the entire servo loop on a machine. Subtle deviations in the data reflect mechanical or electrical problems. A machine's characteristic speed and torque signatures are meaningful diagnostics.

Signal amplitudes, time scales and programmed test conditions are all recorded in the documentation when a new machine is *run off*. Comparing the signature of the new machine with a used and abused machine indicates correct loop operation and adjustment.

12.3.4 Signal Analysis

Instantaneous motor rpm and servo current are directly calculated from the *speed vs. torque* recorder printouts. Continuous power is defined by the area under the data curves. Indications of loop gain, loop overshooting and unstable operation is also documented by the recorded data trends.

Signal analysis is learned at the factory to approve a new machine. Signal analysis in the field relies on noting departures from the approved factory curves. Adjusting the NC parameters and loop settings will directly reflect in the loop speed vs. torque data.

12.3.5 Motor-AC

Most AC motors are not serviceable in the field. Any trouble with the motor or feedback unit requires a complete motor replacement. Call the factory with the motor type and find out the specifications.

12.3.6 Loop Feedback

The style of feedback units fitted to AC motors utilize a host of different signaling schemes that ultimately monitor machine position and speed. Two generic types of positioning systems, the incremental and absolute, use different style encoders. Discussion of feedback diagnostics is split between these two styles.

Of the two position types found in the field, both style encoders are usually built into the motor and difficult to service in the field. The new encoders have detection schemes not found in the old units. Speed and position information is "encoded" in exotic, mysterious detection signals. Around the servo loop there are check points where these special signals are briefly decoded back into simple loop speeds and position signals. The servo application manual defines these points.

Absolute encoder feedback design provides the ability to turn a machine off and on, while still maintaining the same positioning system. These encoders have a memory back-up battery and initialize the control position via serial communication of data packets. There is no useful way to break into that serial data stream signal using the trusty O'scope. It should be noted again that after initial power up, the absolute encoders act very incremental in nature.

With absolute encoders, no zero return is required. Money is saved on some machines by eliminating the "extra" zero return limit switches. However, when absolute positioning systems become accidentally lost or confused those missing switches would sure come in handy to reset the position systems.

12.3.7 Swapping Parts

Moving components from one loop to another is a quick external diagnostic technique. Bad servo-units, NC-servo control cards and motors are routinely identified (review the risks in 12.2.7). Design of the newer machines automatically compensates for replacing servo loop components. The motors, NC-circuit boards and servo amps all tend to be more modular and interchangeable. Servo-units have fewer screws and quickly pull out of the machine. Loop adjustments are spelled out in the factory supplied exchange procedures.

After replacement of one of the servo loop components, increased use of digital circuitry reduces the loop drift common in older, less digital systems.

12.4   The Internal Signals

Computer systems continue the drive towards "user friendly" operations. Some of the "friendly" improvements address internal computer errors and how they are best identified and resolved.

To review, an *internal* diagnostic or *self-diagnostic* is taken right off the machine; perform a few simple keystroke operations and the data flashes right on the displays. Specific internal diagnostics for the servo motion loops are reviewed in the remaining half of this chapter. The internal diagnostics of the spindle loops and computer controllers are treated in the last half of Chapter 13 and 14. All these ideas tend to overlap and reinforce each other.

Old systems just don't have much out in the servo loops that qualify as built-in diagnostics (maybe a few alarm lights or status displays). With progress came the smart AC servos having built-in digital monitors capable of calling up a host of signals like motor current, D/A-command, or core chassis temperature.

On the newer machines, internal diagnostics are continuing to expand. New servo loops are treated with active CNC displays that capture motion trends in "equivalent parameter units." That means the displayed number is keyed directly into a related motion parameter elsewhere on the machine. Diagnostic displays show the answer, and it only needs to be keyed in—incredible stuff!

12.5   DC Servo Loops

There is nothing much to report on the subject of internal diagnostics out in the DC servo loops. The first CNC machines introduced to the market didn't display axis following error, and the signal status diagnostics were almost nonexistent. The servo-units also lacked monitors or alarm lights on the earlier versions.

Technology that followed increasingly supplied internal diagnostics like error pulse, alarm and diagnostic address displays. Improvements were made, adding extra status lights to reflect active loop speed-commands and motor rotating condition.

12.5.1   Computer Numerical Control

The standard operator's manual describes how to find CNC alarms and specific diagnostic addresses in the computer display pages.

Another control feature, the CNC position display, is useful for monitoring the position feedback. When everything is normal, the position display updates as the feedback unit turns.

The loop following error (or *error pulse*) displays the accumulated, unsatisfied position pulses in the servo loop. Error Pulse (EP) also represents the current status of machine motion. A method for calculating loop dynamics uses known loop constants and error pulse running data.

12.5.2   DC Servo Unit

The few lights on the servo units are described in the maintenance manual for the machine. The lights all say something about the loop; maybe it's running, stopped, or in an alarm condition.

Detection circuits inside the servo-unit capture the internal status and alarm conditions. When a light is lit on the older units, one certain way to find its meaning is by referring to the internal schematics for the unit. Finding a copy of these circuits usually requires calling around the network for the old guys who were around when the machine was new. They will have the best answers for antique NC machines.

12.6   AC Servo Loops

Opening cabinets on the machines of today can simultaneously display three or four different styles of loop technology. Using so many different diagnostic systems absolutely requires having the right application and operation manual.

A trade-off is evident as the AC servo loops evolve. The early systems mimicked the loops of the past (a few alarm and status displays and plenty of handy built-in check-pins). Then, the "smart" servo-units arrived with their on-board monitors and stand-alone features. Finally, loop control moved again, this time into the CNC computers, and "slave" servo amplifiers were installed.

Newer machines are currently in a period of specialization of the diagnostic techniques. When several generations of technology co-exist in the same machine, reliance on factory-authorized service inevitably increases.

To combat this trend, the beginnings of standard PC based software controllers have shown up at the machine shows.[2] These efforts towards standardization are driven by market demand.

12.6.1 Computer Numerical Control

Easiest and most popular of the servo diagnostics is the alarm codes displayed in the CNC alarm pages. They often

2   See also the footnote in Section 3.2

lead a servo loop investigation to the right answer in everyday situations. The other, more powerful, internal diagnostics are introduced for the tougher problems—no sense fly hunting with an elephant gun.

Loop error pulse values are often overlooked, but they pack a powerful entourage for field service. The servo motion loops of today are extensively monitored by the CNC computer. Ever-expanding software generated servo diagnostic screens are called up from the operators panel. Some examples of the pages added to the controls include: loop motor currents, high-speed motion compensations, digital auto-tuning and loop-to-loop synchronization settings. Machine shows are an opportunity to see all the newest systems.

12.6.2 AC Servo Unit

"Smart" servos have complete internal display and status monitoring functions. Their expanded displays lead to expanded hours in reading the application manuals.

CNC QuizBox

12.2.1 Sketch the "D/A-command" versus "TG feedback" for the following machining action. Keep the reference to time and voltage the same in all four questions.
   a.) Feed move at 200 in/min, 2 inches long.
   b.) Feed move at 100 in/min, 1 inch long.
   c.) 100% Rapid move at 400 in/min, 40 inches long.
   d.) 25% rapid move, 2 inches long.

13   Spindle Loop

The External Signals 13.1

The Internal Signals 13.4

13.1 The External Spindle Diagnostics

The main spindle unit is fertile ground for connecting up external test equipment. Many service visits are spent parked by one of these units in repeated attempts to find and repair the problem on site rather than resorting to pulling out the entire drive.

Often the culprit is found to be a failure deep within the drive chassis. Removing these shorted components and verifying the control and firing signals restore the main spindle. However, due to the extremely specialized procedures needed for such a main spindle drive service, the job should be passed on to an expert (someone with a good track record who specializes in a specific drive's manufacture and design).

General remarks for this chapter are directed at the more topical systems in use on both the DC and AC spindle drives. DC drives are characterized as rugged, long-time performers. Emphasis here is on keeping them set to factory original. AC drives run hot and fast, the emphasis shifts to evaluating their performance and more advanced diagnostic systems.

13.2 DC Spindle Loops

Early machines run separately excited DC spindle motors with variable-speed, thyrister-controlled drive units. The sheer size of these early drive/motor combinations precludes one man being able to pull them off. A few near antique CNC machines use two separate spindle drives—one runs the motor forward and the second in reverse.

Spindle maintenance was seldom needed in their early working life, maybe an occasional set of motor brushes, followed by a tacho-generator after five years. Rarely did the thyrister "spark plugs" burn out. Some of these units are now twenty years old and still going strong.

Factories originally set up these spindle systems as a matched set. The motor/drive combination was designed, installed and adjusted together. Longevity in the field was the result; however, getting a repaired spindle to run with the same factory reliability has met with varying levels of success.

Over the years, motors are rewound or replaced, and the drive control cards and electronics have gone by the wayside. Even complete drive removal and rebuilds are becoming more routine. Service reliability stems from matching up a new spindle pair electronically, after one or both have undergone a maintenance intervention. A properly adjusted spindle loop again becomes a happy, trouble-free spindle loop.

13.2.1 Signals

A host of well-timed events lead up to a spindle motor rotation. A command is given to *run*, then another to run *forward*, and still another to run at a prescribed *speed* in forward. All these things must arrive at the drive beforehand, otherwise the spindle will not turn.

All the command conditions are checked by the drive before rotation. Once the conditions are approved, the drive begins building the appropriate motor current. The motor quickly responds, coming up to the correct constant speed. At speed, a flat steady *TG* feedback signal returns to the drive. The initial speed request to the spindle system is completed.

The drive holds the motor speed at the commanded level. Occasionally, a minor motor current adjustment is sent to keep the motor speed locked-on and tracking. When the drive achieves the expected performance of a command, a signal is issued to the NC computer confirming speed agreement (SPA). This signal tells the computer the spindle is turning and it's OK to continue moving ahead in the program. Another equally important drive signal, called zero speed (ZSP), is sent when the motor comes to a stop. Both of these spindle status signals are watched by the NC sequence.

The test points for these signals are given in the elementary schematics for the machine. Other useful information like drive status, alarm codes, motor speeds, or loading percentages are found in a copy of the drive's **specific** elementary schematic and trouble-shooting manual.

13.2.2 Alarm Trigger

Older Systems

When a fuse blows on the spindle drive, it protects neighboring components from otherwise certain damage. A service investigation should begin at the first occurrence of such a problem. The events leading up to a problem are recorded. If the cause of the alarm is caught early on, shorting out of the drive's main circuitry may be avoided.

A simple case of worn-out motor brushes leads to both blown motors and blown drive units. Arcing from bad brushes will blow fuses and eventually weld holes in the brush holders or spinning commutator bars. If a fuse is slow to blow, the escaping current spike can short the drive's power semi-conductors. Replacing fuses to blow once again just increases this possibility.

Heavy cutting, bad power and bad brushes will blow spindle fuses and trigger alarms. Capturing a signal during such an alarm event is a powerful service diagnostic. The drive signals are recorded with the alarm event triggering the data. Effectively applied external test equipment allows checking the drive's main input power, motor output, *TG* signal, loading, or any other spindle signal suspected, at the instant of an alarm.

13.2.3 Armature vs. Field

Older Systems

A DC spindle drive sends out dueling currents to the motor. One goes to the armature windings and the second to the motor field windings. Together these currents achieve motor speed and power anywhere within the band of designed running specifications.

Motor *characteristics* describe the relationships between these two currents. Characteristics are carefully designed, written and supplied by the factory. They link the armature and field components across the entire range of motor speeds. Adherence to these factory-prescribed motor characteristics during a service call results in renewed longevity for the drive/motor combination.

When the machine was brand new at the factory, a spindle set-up procedure was performed that joined the drive and motor. Adjustments set the armature voltage and field current to their ideal design curves.

Years later in the field, the field coil windings have slowly broken down. When the wire insulation finally gives way, one-by-one the field wires short, raising the heat and current until smoke comes rolling out. The motor is sent off to a motor shop for a good cleaning and field coil rewind. Once returned it is reconnected to the drive. Since everything initially works, the machine goes back into production. But now, every few months, like clockwork, mysterious spindle problems keep recurring. The machine ran fifteen years without problems, and now the spindle acts up every few months.

Likely the spindle system was disturbed by the new motor windings. They affected the overall motor characteristics by exhibiting a new electrical load to the drive, or to put it another way, the original factory characteristics were lost. A factory field procedure for matching up a spindle drive with a repaired motor is needed. Motor type and drive type will summon up the correct set of characteristic curves and potentiometer settings from the factory.

Running the spindle in steps across the range of spindle speeds generates the necessary adjustment data points. In all, three signals are monitored while iterative adjustments arrive at the final drive settings. Many times, to simplify the equipment connections, the signals for field current and motor speed are derived from handy voltage signals present at the drive. The OEM can offer the specific procedural recommendations.

13.2.4 Signal Analysis

Older service documents at the factories contain actual photos of oscilloscope screens. These documents give many examples of the analog signals running around inside spindle loops. Performing electronic signal analysis on old DC drives using reference photographs is wonderful when such pictures exist, but most times they do not.

The host of electronic signal data captured and displayed on test equipment must stand up to a critical review. Missing signals, clipped signals, noisy signals and over-driven or saturated signals are all confusing and should send up a red flag. Careful comparisons with known good signals helps establish certainty in otherwise unknown surroundings.

Signal comparisons performed on the same circuit board or between like boards is called *signature analysis*. Equipment for checking a signature is found in the shops dedicated to board level repairs. The usefulness of this equipment is determined by the system price tag and operators ingenuity.

13.2.5 Motor-DC

The size of a direct current spindle motor adds challenge to a repair. All items that can be checked with the motor in place are investigated before ever thinking of a complete motor exchange. The motor brushes, speed generator and main incoming power are addressed. The armiture is checked with both a *megger* and a simple back *emf* generator test.

If everything looks good and the trouble persists, the remaining service options get expensive. In simple terms, it's either the drive or the motor. Shipping away both units to be tested is an option. Ordering replacements is an option (if any spares are actually available). Buying a new retrofit replacement unit is the long term solution.

Someone needs to bite the bullet and make a tough decision. Bring in more people, parts and opinions. Meanwhile, the downtime mounts while waiting for a miracle. No matter how it's sliced, the options are expensive. Big, old, intermittent spindle problems darken the doorway of businesses around the network. The retrofit option is gaining favor.

13.2.6 Loop Feedback

A speed generator is mounted on the motor. Workings of the generator are found behind a sheet metal inspection cover. As always, some specific advice is best faxed over before inspecting the unit. Every OEM fitted their own version of a speed detector. The detector discussed here is called a brushed DC tacho-generator.

A worn generator exhibits a build-up of black carbon dust inside the sealed cover—any build-up is blown away with clean, dry compressed air. With the dust gone, the generator's tiny brush commutator is inspected for evidence of grooving. Old units develop deep grooves from the action of the miniature brushes.

The entire commutator is removed with a small gear puller or blind-threaded stud arrangement. If any of the fine braided generator wires are accidentally nicked, the unit is ruined. For minor wear, the grooves in the commutator are turned down. The factory can supply a set of replacement brushes, or a complete spare if the unit is almost worn out.

Electrical generator function is checked by running the spindle at constant rpm while measuring the signal ripple with an oscilloscope. Before cleaning the generator, the signal ripple is first measured. The ripple is measured again after cleaning to judge the lasting operation of the unit.

Main motor current is sent in response to this feedback signal. A smooth, tight generator signal results in a smooth, tight motor current. When a dirty speed generator signal is replaced or remedied, the motor operation smoothes out dramatically.

13.2.7 Swapping Units

Older Systems

Some shops keep older machines running in pairs to help out with troubleshooting. News of a machine being repaired by a part from another machine is not surprising. When everything can return back to original, without causing a new problem, the gamble has paid off.

Moving parts around is risky. As a rule, if spare parts are needed for swapping, order replacements. No sense needlessly damaging two machines. In practice, replacing the entire spindle drive unit is a better idea than swapping around individual components like PC-Boards.

After someone else tampers with a DC spindle loop, the next visiting service engineer is a little anxious. They know how one faulty connection can immediately send the giant spindle motor into a dangerous runaway condition at power-up. All the drive wiring and signals must be checked before power to avoid unexpected damages.

When a freshly reassembled spindle system is powered up (using new, rebuilt, or borrowed parts), an alert hand is posted away from the machine at the main tap switch for the machine. If something looks abnormal, the machine is quickly clicked off. Due to the extremely specialized procedures needed for such a main spindle drive service, the job should be passed on to an expert (someone with a good track record who specializes in a specific drive's manufacture and design).

Very few people allow swapping parts for an excellent reason—the risk of putting the second machine down. Only the shop supervisor or owner has the authority to direct the swapping out of a good machine to get a down machine going. Most would prefer to wait rather than to risk a bigger disaster.

13.3   AC Spindle Loops

Newer Systems

A modern machine has alternating current (AC) motors and drive-units. A concern surrounding these AC spindle drive designs is the dangerous bank of 300 Vdc capacitors that remain charged long after the drive power is switched off. This fact alone recommends passing on the internal repair jobs to those having good track records with fixing a manufacturer's specific spindle type. Only they will have the extremely specialized procedures necessary to work inside a main spindle drive.

After a drive is smoked, skilled testing with voltmeters will find shorted and damaged power semiconductors. Then, using oscilloscopes, all the control firing signals are checked before replacing the shorted transistors and returning the big and dangerous drive currents.

The discussions here will only center on two facets of AC spindle loops: the dynamic performance of these drives and the many generations of drive improvements to watch for in the field. Observing spindle loop status and alarm signals with diagnostic equipment will generally reinforce the understanding of both these points.

13.3.1 Signals

Newer Systems

Factories supply designated checkpoints for connecting external test equipment. Standard access is provided for motor speed and motor current signals.

The latest AC spindle drives include dozens of signals devoted to input and output control. Fortunately, these newer drives also offer built-in display monitors to read all these status signals. Checking a drive's status with external test equipment applies mainly to the older units, without the convenient status monitors.

217

13.3.2 Alarm Trigger

A remote gang of alarm lights signal trouble on the first generation of AC spindle drives. Since then, new on-board computers have arrived that capture and display spindle alarms in alpha-numeric *alarm code* formats.

These new alarm codes are fairly meaningless without the corresponding factory alarm lists printed in the spindle manual. These hefty manuals explain all the parameters contained inside the drive's computer memory (and a host of other internal type diagnostics reviewed further in section 13.6.2).

An alarm on the drive is communicated to the main NC system by alarm signals. The same alarm signal sent to inform the NC can also trigger a data recorder. Newer generation units use transistor-driven alarm signals instead of the conventional reed relay contacts found commonly on the older units. Either style signal effectively triggers a data trace during investigations.

13.3.3 Speed vs. Torque

Capturing a set of standard drive signals will fully document a spindle's dynamic operation. One standard is a combined trace of motor speed versus motor torque captured during a test program run-off. The factory often includes these spindle traces under specified test program conditions as part of a new machine's documentation.

A set of these traces captured over one part cycle can take an intermittent spindle problem up to the next level. When alarms are involved, the data leading up to the alarm trigger is recorded.

13.3.4 Signal Analysis

A data trace of spindle torque versus speed over a periodic test cycle allows calculation of the continuous output power of the drive. The shape of the data also suggests effective drive or program adjustments.

During a rapid acceleration of the spindle motor, the torque shoots up to the top limit set for a spindle drive. This torque limiting (TLIM) level is easily seen in the trace data and sets a convenient calibration to the data. Looking up the equivalent ratings at *TLIM* suggests the continuous drive power of an application. New hand-held scopes simplify such power calculations by using built-in signal analysis routines. Either way, integrating the trace data for the area under the curve gives the power.

An excessive up and down part cycle never lets the spindle rest. Inspecting the captured speed vs. torque data trace highlights this point—it shows how each change in speed pegs the torque (and current) to the limit. As this rapid up-and-down cycle increases, the continuous level of spindle power eventually climbs above the rated power, and intermittent spindle alarms result.

Writing a speed clamp in the program clips off the unnecessary up-and-down speeds, without generally affecting the part cycle time. In this way, a speed clamp reduces the continuous spindle power—sometimes a simple clamp drops the running power back within the factory specification.

13.3.5   Motor-AC

AC induction motors have the advantage of maintenance-free reliability. Subsequently, few techniques for checking a troubled motor are needed. Two minor observations are mentioned in the material that would qualify as diagnostics.

Simple observations can lead to the most interesting diagnostic features.

A loud noise from a running spindle motor has either a mechanical or electrical source. With all the commotion, it's hard to pinpoint a bad bearing or jumpy motor current as the cause.

If the motor current is suddenly cut off, the motor will coast to a stop. Remaining noise during the coast is the identifiable diagnostic. If the motor noise should suddenly disappear while coasting, a jumpy drive current caused the problem. On the other hand, a howling pitched noise while winding down is a sign of a mechanical origin. Check with the drive OEM for suggestions on safely coasting a motor down from speed. Some recommend cutting off the main breaker, others suggest using a drive's own base-blocking alarm event.

Electrical resistance testing of a three-phase induction motor normally shows it shorted phase-to-phase, and open from phase-to-chassis.

13.3.6 Loop Feedback

Newer Systems

Closed speed control relies on the feedback from the motor. Each motor manufacturer provides a feedback style compatible with the other loop control electronics. Spindle axis systems offering servo mode *orientation* adds additional influence in the maker's choice of feedback.

A rugged and reliable resolver type feedback was popular on the early AC spindle systems. Only a broken resolver winding or loose rotor caused speed detection malfunctions. An oscilloscope can readily capture resolver excitation signals for whatever limited diagnostic purpose they offer. Designers of resolver feedback deserve applause for their maintenance-free longevity.

New, high technology pulse encoder feedback units offer the combination of speed, position and marker data in a pulse train format. A spindle pulse encoder failure generally requires a complete motor replacement. However, there are some motor designs which allow field encoder replacement. Check with the maker for the latest repair recommendations.

A few machines have remote feedback units mounted off and away from the motor in an area where the indirect spindle rotation is picked off. A motion linkage is built up with either belts, gears, or a direct drive system.

13.3.7 Swapping Parts

Newer Systems

Complete AC drives and motors are relatively compact and self-contained compared to their DC style predecessors. Complete replacement procedures are offered by the factory service centers around the world.

Complete exchange of the older drives requires a transfer of the solder-on components, screw potentiometers, *DIP* selection switches and plug-in style jumpers to the new unit. A faxed procedure from the OEM is strongly recommended. The motors seldom fail unless the bearings or shaft is physically shot.

Exchanging the newer, space-age spindle drives demands attention to EPROM software chips and long listings of user parameter data. Such information is either keyed into the drive memory, or sent over a serial RS232 connection. An exchange procedure is again required from the OEM.

Any work inside a drive unit is discouraged. Call the factory service people, dealer, or any qualified service company having experience with a drive's specific design, layout, and OEM-issued repair procedures.

13.4 The Internal Spindle Diagnostics

Over successive generations of AC spindle drives, the growing requirement for communicating spindle drive status with the main NC computer has encouraged the expansion of spindle loop diagnostic displays. With the advent of on-board drive computers in recent years, the diagnostic displays devoted to spindle loops have really taken off.

Early vector controlled inverters had the familiar green status LEDs seen before on the DC spindle units. Service for older units would involve hours connecting up test equipment to see a drive's status or command signals. This job is now finished in just minutes, using the new expanded on-board diagnostics.

13.5   DC Spindle Loops

With the exception of a few indicator lights mounted on the drive, older DC spindles have limited internal diagnostic function. The NC computer supplies the majority of resources for checking a drive's status by using the alarm and diagnostic pages in the control. Physically connecting voltmeters and other test equipment to drive-signal wiring is common during field work on the old DC spindle systems. This involves measuring the voltage of each status signal using the check points given in the drive schematics.

13.5.1 Loop Diagnostics (NC-Side)

Ones and zeros recorded from the NC diagnostic displays should exactly mirror the direct voltmeter results off the drive. When signal data doesn't match display data, a problem is isolated.

Spindle loop signals like *RUN* and *FWD* are reflected in the CNC diagnostic lists. These signals are considered computer outputs (or drive inputs). Likewise, loop signals like *SPA* and *ZSP* are also found in the diagnostic lists, but this time as control input signals.

13.5.2 Spindle Drive-DC

By today's standard, the supply of internal diagnostics found on the old spindle drive systems is sparse to none. A few green LEDs on the front circuit cards for zero speed, speed agree, and maybe a few small red LEDs to inform of an alarm condition on the drive.

13.6   AC Spindle Loops

Transition to digital circuits and Digital Signal Processing (DSP) fueled the diagnostic revolution. Single, seven segment displays were followed-up by the multi-digit, key-button controls of today.

The diagnostic features showing up on new AC spindle drives offer quick drive assessment and adjustment without the need for bulky external test equipment. The powerful central drive computers of today contain on-board parameters, software control, expanded I/O-Status listings and detailed alarm protection.

Factory documentation is indispensable with the newer systems. The little digital monitors are extremely confusing without the specific written directions. Only the spindle troubleshooting manual offers the detailed adjustment procedures, data addresses and keyboard routines needed for a drive service. Drive investigations require the correct version drive manual.

13.6.1 Loop Diagnostics (NC-Side)   Newer Systems

To complement the control of *solid tapping* and *synchronous* running of two independent spindles, an error pulse display of spindle operation was incorporated. Accurate adjustment of both these new functions requires such error pulse data from the spindle loops.

Control-side monitor and display of the actual spindle speed was available on some older machines, but has now become standard.

Orientation of the spindle position, or servo-mode, is offered in several designs. The most widely seen system is the familiar magnet and sensor system. Another system uses a pulse encoder feedback signal with the pulse counting control-led by either drive or NC-side circuitry.

13.6.2   Spindle Drives-AC

Newer Systems

Diagnostic monitors on today's drives continue to im-prove. Insightful new techniques for diagnostic-based field solutions continue to surface.

The new drives do not require potentiometers on the cir-cuit boards—all adjustment is contained in the software driven parameter settings. New drives are treated like a computer; the drive control data is saved before starting a repair. With the data saved, a return to original condition is assured. More data is inside these new spindle drives than the earliest NC comput-ers.

Internal spindle diagnostics cover three key functions: signal status display, application setting adjustment and alarm indication.

Status includes motor temperature, bus voltage, instanta-neous power output and scaled or actual motor rpm. Complete access and display of each I/O communication signal is stand-ard in the status monitors.

Application settings are carefully described in the drive's application manual. The OEM calculates the correct list of ad-justments for each type of machine they sell.

The instructions for fixing a drive problem are improved. Alarm lists now include expanded reasoning and corrective ac-tions for many possible spindle fault conditions. The sequence of events leading up a spindle problem may even be stored in a drive's alarm history memory for a later on-site retrieval.

The convenience offered by these new spindle systems is reflected by a quick example of the steps taken to correct a speed drift found inside a spindle loop.

On old spindle systems, a voltmeter is hard-wired into the D/A-command circuit. The meter is needed to monitor the re-action of a potentiometer adjustment dialed-up on the NC-side spindle speed command board. A second, similar procedure is again carried out on the spindle-side of the circuit to complete the overall adjustment.

On new drives, a unit value of D/A drift is shown right on the drive's internal status display. The displayed unit value is merely keyed into another drive parameter. This correction removes the command drift.

CNC QuizBox

13.2.3 Calculate the total resistance of a spindle motor field winding. The copper wire diameter is 2 mm; length is 10m.
   a.) At 68 °F.
   b.) At 20 C°.
   c.) At 100 °F.

13.3 Calculate the rectified bus voltage from an AC signal source represented mathematically by $v(t)=V_m sin(wt)$ Volts.

13.3.4 Find the average power of the periodic data trace captured below into a 1 ohm purely resistive load.

14   Computer Numerical Control

14.1   The External CNC Diagnostics

The subject of this chapter is the *internal* and *external* diagnostics of the CNC computer. Together, these two techniques can document and solve a wide range of problems associated with machine tools. The distinction between a control or machine problem is often made using the CNC diagnostic systems.

In the field, diagnosing the computer's internal function relies on the *internal* diagnostics supplied by the maintenance pages, parameters and switch monitors in the control (see Section 14.8). The common *external* diagnostics for field service include visual burn marks, contamination, machine vibration and board swapping.

The outward appearance of a standard CNC computer exhibits very limited access to basic test equipment. However, lurking beneath the surface of all the computer boards is a specially prepared array of highly specific factory test systems for use when building CNC simulators, carrying out board-level testing or other factory component repair. The domestic and overseas repair centers invest heavily in these automatic test systems to quickly analyze and repair the circuit boards damaged in the field. A few after-market companies are offering to reverse engineer and repair damaged boards.

Topics of this nature, while interesting, exceed the scope and training necessary for field level repairs. Call on the network for circuit board testing, repairs or specific diagnostic set-ups.

14.2   Visual Effects

Sometimes, the best repair suspicions are raised by just stepping back and taking a good, long look. A connector found unplugged half-way is noteworthy. Burn marks and smoke trails raise a curious eyebrow. Stray metal chips scattered around the CPU rack is suspicious.

A close inspection of a printed circuit board shows that a thin substrate was photographed, etched, drilled and mounted with a variety of miniature components. A closer inspection reveals each component has a board location and printed symbol. Protruding alloy plated check-pins attract the greatest service interest. Older systems offer abundant check-pins and check points; there are few to none on the newest digital systems.

14.3   Vibration

A machine's constant companion while cutting metal is vibration. From a service viewpoint, vibration affects all the electrical components and wiring connections on the machine. Longevity and reliability depend on isolating and eliminating excessive vibration before it causes complete electrical failures, or those dreaded, unexplained machine glitches.

A good preventative maintenance program considers machine vibration. If the machine is not properly leveled on the floor, an undamped, rocking vibration occurs. An inadequate concrete slab under the machine serves to amplify the problem. Underneath the machine casting are leveling pads—if they are found loose, the machine is settling into the shop foundation. The location of the machine on the floor needs checking followed by a proper leveling of the entire machine.

A machine's general application often suggests the source of excessive vibrations. Running hex-shaped material in a bar feeder set out of alignment will set up an excessive vibration. Any spinning shaft, set out of alignment with the driven equipment, will generate harmful vibrations. Cutting on rough castings, where a tool hits only into the material on a few points, causes excessive, and often unacceptable levels of vibration.

A vibration related problem is addressed by finding a way to turn it off and on. Reducing spindle vibration will affect the entire machine, so problems related to spindle vibration will temporarily disappear while using a *speed clamp*. Clamping down maximum spindle speeds (with a G50 S-code) quickly separates a suspected vibration problem.

A machine sitting idle when an alarm generates would tend to rule out vibration as the culprit. Occasionally, a field problem with vibration is identified by gently tugging and tapping on the cables, *ice-cube relays* and PC-Boards. Sensors are available that can capture vibration levels with standard external test equipment. To be effective, these type of measurements are repeated over several bands of vibration frequency.

A discussion of machine tool vibration wouldn't be complete without mentioning *gain* and its role in the electronic control loop. For this simplified discussion, gain is defined as a rate of correction or comparison established within the two closed motion loops. As the gain of a servo loop increases, vibration is an eventual result. Common symbols assigned to a machine's loop position- and velocity-gain are K_p and K_v. The exact conditions for closed loop oscillation and vibration is a keen subject in the factory analysis of a new machine tool.

Construction of the machine yields limits for overall K_p and K_v. A flimsy machine has different limits than a solidly built unit. Machine and control makers are very cognizant of the gain calculations between competing machines. Factory-chosen loop settings serve to eliminate loop gain related vibration. Tool and table vibration is a problem to be overcome before a good metal cutting machine can properly operate.[1]

14.4   Power At Boards-DC

One of the few signals observable on every NC computer is the DC power. An adjustment procedure (must come from the OEM) identifies all the handy check points and proper ad-

---

1   See also footnotes in Section 3.6 and 6.2.

justment potentiometers. Checking directly at a power supply
unit shows a higher meter reading than checking the same circuits down the line at the PC-Boards. Haphazard adjustments
to the DC levels traveling around the machine can blow out the
early circuits while trying to bring up the level some ten-feet
down the line.

14.5   Contamination

A keen eye is kept on the lookout for foreign substances
inside the machine cabinet—oil condensation, metal dust and
shavings are all bad signs. Cleaning up a mess is a gamble, it
may suddenly resolve the strangest of problems or put down
what was a perfectly running machine.[2]

A cabinet mess is cleaned one piece at a time. A small
piece is cleaned then the machine checked out for new problems. After the machine checks out, on to another piece of the
mess. Cleaning too many areas at once and clicking on the
power is a prescription for disaster. Small steps help identify
what cleaning caused what new problem.

For big messes electronic de-greasers are used. These
will dry without residue but can inadvertently flush dirt from
one board onto another. Good ventilation while cleaning
avoids getting dizzy from breathing fumes and also lessens the
risk of explosions. Fumes from the cleaners will explode from
a single spark, so there is no cleaning with the power on! After
a mess is cleaned up, the source is identified and sealed up.

14.6   RS232 (Diagnostics)

In earlier sections (7.9 and 9.2.3), the standard RS232 interface operation was briefly discussed and outside references
given. Now the interface itself is tested and applied to diagnosing problems on a machine.

2 See also the cautions in Section 8.12

The first step in testing a dead interface is trying it out. If sending "in" and "out" hesitates to work, the obvious items are swapped around: try two different host computers, different data cables or two entirely different NC machines. Most shops have many machines and many communication systems for swapping. A visiting service company will also carry their own communication equipment. One of those tests will isolate the problem. A dedicated RS232 *break out box* is another option for locating communication troubles.

Communication networks which are hard wired into the interface are put at risk during a lightning storm. If a lightning surge is picked up by the communication cable everything connected in its path is damaged. Unhooking or optically isolating interface connections is always good idea, and in the areas prone to thunderstorms, it's an absolute necessity. Technically, such damage is considered an "Act of God" and excluded from a control warranty.

After the damage is done the standard send and receive circuits (chips like SN75188s and SN75189s)[3] are visibly checked for obvious pops in the case. If these chips are plugged in sockets, a fresh set is tried. If they are found soldered in place, a fresh communication PC-Board is ordered. The OEM and dealer are well versed in the best repair options. If the damaged board contains the only communication interface, saving all the control data is impossible. Hours can pass keying in all the back-up data by hand.

A standard interface has a data *collation* or verify feature that compares a file stored in memory with a file sent in from outside. If these two files have differences, the collation fails, and an error message is displayed. When the two files are identical, the test passes. Using "verify" puts to rest any wild claims of missing, or "mysteriously changed" control data.

---

3 Texas Instruments Data Book for Standard TTL Components.

The verify function also guarantees that a saved file will indeed load into the control if needed later. Nothing is worse than saving all the NC-data, regenerating the control and then finding out all those big saved files are glitched and won't go in. Verifying offers piece of mind. It's also effective for checking program files and offset files against the shop's certified file to rule out any surprise changes or edits.

A high-speed RS232 communication option is quickly tested by simply slowing the high speed port down to the rate of a standard port. After a few comparisons both ports are checked, then the high speed data rates are pushed back up.

14.7   Swapping Parts

A "swap-a-neer" is an engineer who begins swapping parts at the start. And certainly, replacing parts in the NC control computer is possible with the OEM's help. However, before ordering up parts, getting procedures and dismantling the machine, collect a full description of the original problem and elicit a second, possibly third opinion. A shotgun swap festival is one real quick way to lose the original reliability of a CNC-tool, it also puts a service technician in needless danger.[4]

The primary importance of getting the machine running should balance with finding a full explanation of the trouble. Once parts are taken from the machine, the original condition is lost.

14.8   The Internal CNC Diagnostics

Years of unending operation can pass without a major service call, such is the quality of these modern tools. When there is trouble, the most powerful resource is the built-in diagnostics from the machine's own operator's panel display.

4   See also the cautions given in Section 5.4.11, Section 6.4.5 and Section 13.2.7

From the panel, the status of *every* switch, lamp and coil is observed in the input/output diagnostic addresses. A further listing of generic alarm codes is also given at the panel. With experience, these internal CNC diagnostics connect machine problems with their definite, underlying cause.

Observing the diagnostics flashing over the operator display is certainly the cleanest way to fix a machine. The risk of getting shocked or somehow damaging the machine is reduced, since nothing is touched and only data is observed and written down. A careful interpretation of the captured signal data using alarm lists, ladder diagrams and control signal assignments arrive at the answer.

In some cases, putting external test equipment to the circuit in question verifies the data received from the panel display. This practice of verifying diagnostics, comparing the internal signals with the external signals and vice-versa, ensures that the schematics and diagnostic lists are in agreement. Sometimes a wrong version drawing is found at the machine which misleads the repair effort.

14.9   Finding the Diagnostics

The control builder provided internal diagnostics to diagnose problems in the field. The body of internal diagnostics are displayed at the operator panel. Specific subsystems like the servos and spindle drives can also provide limited on-board diagnostics and are addressed at the end of chapters twelve and thirteen.

A machine needs to be put into *maintenance mode* to access all the internal diagnostics. While most of the diagnostics are available in standard running mode, certain operations will only come up after the control *system switches* are set for maintenance mode. When a special color or graphic computer is present, additional maintenance pages are added to maintain the second, built-in computer platform. OEMs often supply access information in the procedures written for a specific control or specific machine.

A cursory explanation of how alarm and diagnostic information is observed is given in the operator's manual. All the internal diagnostics supplied for the machine are cross-referenced and inter-related with the lists of alarm codes, signal addresses and wiring schematics. Those fixing the machine use these lists to determine the exact cause and effect of a problem.

14.10   Operator's Panel

The operator runs the machine while standing at the operator's panel. The top half of the panel is devoted to the control display and keyboard and the bottom half holds the push buttons, toggle switches and small indicator lights. Panel use and operation is explained in the machine manuals.

Bringing in programs, moving the machine around, or reading off the current machine position is all done from the main operator's (OP) panel. When something goes wrong, the OP panel displays the news.

14.11   Sampling the Diagnostics

Understanding how the internal diagnostics work is best described by just picking out some sample signals from a machine. A good sample would be a switch that lights up when it's pressed, maybe the Single Block (SBLK) switch or coolant-on push button. According to the diagnostic list, the switch (input) and lamp (output) addresses are first identified, then those two diagnostic addresses are simply called up on the operator's display. (Note: Always have a qualified operator assist with any actual machine maneuvers or operations.) When the switch is flipped on and off, the switch and light actions are mirrored in the diagnostic data. If necessary the signal addresses can be followed further, all the way deep into the control computer.

The SBLK switch signal passes through the PLC Sequence and into the heart of the control. The control internally activates the single-step function of the SBLK and then launches (onto another signal address) the light output signal, for the return journey through the PLC and on out to the SBLK bulb.

At this point, imagine that two *levels* of internal diagnostics exist. At *level-one* are those signals hard-wired on the machine from the I/O-Board. At *level-two* are the internal signals, where the PLC-Sequence meets up with the NC computer. Usually, level-one signals are best for maintenance investigations. Level-two signals are avoided because they quickly become very complicated and misleading in their analysis.

Level-one signals are readily verified by placing a voltmeter across the physical signal wires leaving the computer I/O-Boards. When toggling the SBLK switch, the voltage across the switch contacts changes from 24 Vdc (open contacts) to 0 Vdc (closed contacts). Miraculously, the diagnostic address for the SBLK reflects these actions by changing from a logic zero (switch open) to a logic one (switch closed).

14.12 Lamps and Switches

"Real world" signals are monitored by *level-one* diagnostics. A light coming on can be viewed in two places: first in the diagnostic address, and second at the bulb. The same holds for the switches—first the contact closes, and then the switch diagnostic reflects the change.

Because of the vast number of signals needed to control a machine, the input "switches" and output "lamps" are separated by a *prefix* designation used in the diagnostic lists. A set of prefixes are assigned to help find and distinguish all the level-one and level-two signals. Understanding the signal prefixes of a particular OEM quickly sorts out their long lists of mysterious signal addresses.

During service, the prefix for "hard" switches (that are mounted out on the machine) is much more useful than the prefix for, say, the level-two signals deep inside the NC computer. Prefix assignments are printed in the maintenance manuals, and with experience, are easily searched out from a machine's PLC ladder diagram.

To decode a prefix scheme requires following the SBLK signal all the way around in the ladder diagram. Either a common letter or number scheme will distinguish the many groups of diagnostics from each other. The factory service centers can be called on to explain their particular prefix assignments.

14.13   Deep Inside Signals

The CNC processes the *level-two* signals. The real world, level-one signals are converted into deep-inside, level-two signals before reaching the NC computer. The computer flourishes in this level-two world, safely shielded away from the level-one signals.

The nature and description of these deep-inside, level-two signals is very complicated. Because the processes take place so deep inside the control, few level-two problems are worth pursuing from a maintenance standpoint. The chances of solving a problem are much better in the real world, level-one arena. However, there are two brief exceptions worth discussing.

First, there is a family of level-two signals called *direct-in* signals, similar in many respects to level-one. Direct-in signals are those used for touch probes and other high-speed switching signals where the time needed to process a signal through normal I/O channels is too long. The direct-in signals are sent directly into the heart of the computer to save critical signal processing time.

A second example occurs when a new machine is rushed to market with a *bug* in the NC software that involves the level-two, deep-inside signals. The final solution to this kind of problem will come from the control builder in the form of a new NC main software update.

Considering a suspected level-two problem without knowing the connection to the level-one signals gives premature results. Problems in the field are first documented using the level-one real world signals, after that a more detailed investigation at the factory will take it on up to level two.

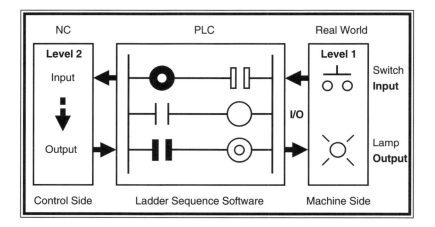

Figure 14.1 The Signal Loop

14.14 Ladder Sequence

Each machine type is somehow a little different. When the machine maker originally decided to install a certain type of computer controller, the machine characteristics were all accommodated in the *ladder sequence*. The *ladder diagram* links together the infamous level-one and level-two diagnostics. The ladder sequence diagram is typically the longest maintenance document supplied with the machine.

After the long process of writing the machine sequence, the software information is often burned into an EPROM and plugged into the control sequencer board. The version of PLC software plugged into the sequencer should match the version printed on the ladder diagram. Reading a mismatched ladder diagram is a common problem heard from the field.

Like the SBLK example from before, following a known signal from switch contact all the way into the control and back out to a light output requires a number of cross references. The ladder diagram, machine schematics and elementary control schematics are all intimately involved.

The different type of signals in the ladder diagram are assigned visual *symbols* to ease the analysis. There are level-one and level-two symbols, symbols for *sequence timers* and symbols for inputs and outputs. Symbols can also indicate whether a signal is *active high* or *active low*. Each type of diagnostic address is assigned a corresponding symbol in the ladder diagram.

Certain parameter addresses called machine or sequence parameters are set aside for exclusive use in the sequence software. As an example, a *toggle* type sequence parameter can interlock parts of the ladder sequence, a *value* type sequence parameter can give timer control to the sequence.

Sequence timers are a good example of *macro ladder instructions*, other macro instructions include decoders, turret controllers and bit instructions. The timer macro instructions build a window of time during which a signal can pass. If a signal is late in arriving, the timer will *time out*, effectively blocking its passage through the sequence. Such a ladder time out can generate a *sequence alarm*. In this way, an alarm is detected when machine events happen out of their proper sequence.

Two types of timers are found in the ladder. *Fixed timers* are of a fixed duration and permanently burned into the software—only by changing sequence EPROM's can they be adjusted. The second type of timers are the *adjustable timer* which are adjusted by increasing or decreasing a related sequence parameter in the control.

For instance, the signals from a tool changer must arrive within a set period of time. Too long or too early of a delay may indicate a mechanical problem has developed. When the ladder sequence times out, the machine shuts down. Until a tool changer's sequence of signals is restored to its proper speed, the same sequence alarm will occur every time during a tool change.

Reading a ladder diagram relies on using the tiny cross-referencing numbers scribbled throughout the diagrams. These small numbers are placed next to each coil symbol in the diagram in a fractional style format of page number over line number.

The number of distinct diagnostic addresses listed in the sequence ladder for a lathe will number in the hundreds, sometimes in the thousands for a machining center. They are all inter-related by using the signal prefixes, ladder symbols and cross-referencing numbers. Unfortunately, most control builders have different symbols and conventions in their ladder drawings. The best way to decipher the operation of the sequence is to contact the control builder.

New options for a machine occasionally require new PLC ladder software. By plugging in a new EPROM (or uploading data files) some improvement to the machine sequence is installed. Some control builders offer ladder editing systems that support modification of the ladder sequence software from the machine operator panel. Others offer proprietary off-line editing systems for making software changes on a desktop computer, finally burning a new EPROM for use in the control.

The use of the old style ladder diagrams during a machine tool service is difficult, a new approach that relies on flow chart programming is rapidly gaining favor by the major control builders.[5] Following a sequence of events using a flow chart is straightforward compared to tracing through the many logic conditions kept in the interrelated rungs of a ladder diagram.

5 Steeplechase Handbook, A Practical Guide to PC-Based Control & Flow Chart Programming, Steeplechase Software Inc.,1997.

14.15   Ladder Safety Interlocks

Every machine type has a set of custom alarm interlocks supplied by the machine builder. These special safety interlocks are built using clever interconnections expressly written into the PLC ladder sequence. For this reason they are called the sequence interlocks and alarms. The NC computer immediately stops the machine when sequence alarms are detected.

These important safety interlock functions are highlighted by cross-referencing a few signals in the ladder diagram. Consider a specific interlock called the *cycle start interlock*. In this interlock, a program cycle is only permitted after everything on the machine is checked and found to be in "run" condition. If the status of the machine changes, say, the machine's lubrication oil tank becomes low, or the material feeder for the machine becomes empty, the interlock is activated and the machine stops. If the spinning part is not clamped securely, the automatic operation stops again until the run conditions are correct. In the previous three examples, adding oil, loading up the bar feeder, or clamping the part securely would clear off the cycle start interlock.

These and many other interlocks are creatively written into the sequence and are tough to identify using the ladder diagram. In some cases, the ladder will include helpful maintenance messages to speed up interlock investigations. The heavily interlocked systems, like cycle-start, feed-hold, pallet changing and the electro-mechanical magazines for tool changing, offer the toughest tests for interpreting the ladder diagram on service.

14.16   What's the Story?

The toughest problems are put to bed by collecting the sequenced status of the diagnostics. A clear chart of progressive steps taken by the involved diagnostics is proposed and tested. With success, the diagnostic documentation will interpret the entire hidden story.

14.17   Parameters Effect

Meaningful and specific changes to the parameters can act as powerful diagnostics. Each assigned parameter does something: changing one parameter slows the machine down, another slows down just the RS232 baud rate, and another slows down a sequence timer. A machine is initially set up with a *basic list* of factory calculated parameters. Any changes from this *set-up* list technically puts the machine system out of factory adjustment. Changing the right parameter can fix a machine, or it can increase the frequency of a machine's problem.

Two good rules to follow for changing parameters are: (1) recording the original parameters before changing anything, and (2) knowing in advance the full and exact function a change is going to cause. All the features controlled by a parameter are not always so obvious. *User* parameter addresses and their functions are listed in the operator's manual—all the others need accurate advice and help before changing. Accidentally changing the wrong address can quickly crash a system, electronically or mechanically.

Whenever the parameters are lost from memory the most current back-up list is used. Sometimes, a spare back-up list is kept at the dealer and other such sources. Parameters are reloaded using the OEM's complete, written reboot (regeneration) procedure. If no back-up is available, the machine's basic set-up list is loaded, and a few days or weeks are spent getting the machine back in shape.

Temporarily setting a known parameter to an intermediate test value will help find a problem. Suppose a backlash parameter is zeroed out. At that time, a test cutting from the machine with zero compensation will reflect the actual condition of machine mechanicals. Other machine functions are isolated for testing by knowingly increasing, decreasing, or setting to zero the related function parameter(s).

Every time a machine problem disappears by changing a parameter begs the question, "What's keeping the machine from running at its correct factory settings?" Without answering that question, a problem isn't completely fixed. Leaving a parameter at the wrong value can lead to some very surprising and expensive damage down the road. The set-up list worked on a new machine, so when the machine is right, the set-up list will work once again. Keep the machine set to factory specifications.

14.18   Alarms

Protection to the machine and its operators is provided in large part by the built-in alarm systems. When an alarm is captured, the system is halted until the cause is identified and cleared.

Alarms are generated by several methods. The detection methods are usually grouped into either a hardware or software source. An example of hardware alarms is a blown fuse. Software alarms come from software routines stored inside the control computer and inside the sequence software.

14.19   Alarms from Hardware

The source of a hardware alarm is a hard switch. A contact somewhere on the machine opened, and an alarm was captured. If reset does not clear the alarm, the source must still be present. Locate the contact source, and the cause is found.

A hard alarm is triggered when a switch contact opens a normally closed alarm circuit. This signal change is detected at the I/O-Board and sent into the NC computer. The computer connects the signal address with the alarm name and stops the machine to display the event. The alarm remains active until a *reset* is performed.

Hard alarms are reset by closing the responsible alarm contacts, either by replacing a blown fuse, resetting a tripped circuit breaker, or re-engaging a thermal overload on the machine. The open contacts are located using either the alarm description, the diagnostic address, or just tracing the circuit down with the schematics. If necessary, an alarm status and contact status are matched up with a probing voltmeter.

If pressing "reset" at the operator's panel clears the root alarm, then whatever hard contact that had opened has since closed. The hard alarm trigger was actually *latched* by the computer software. Pressing reset *rewound* the sequence, clearing off all the alarm latches held in software. By the time the "soft" reset button was pushed, the "hard" sensor contact had returned to a closed position. Maybe an overheated temperature sensor initially triggered the alarm, but had since cooled back down and closed.

Imagine how a loose connection could generate a real pesky, intermittent alarm. Any vibration would cause a hard alarm circuit to open for a split second, just long enough to be captured by the computer's alarm sequence. Going out to check the alarm circuit shows it's always closed; it was a *false* alarm caused by a faulty alarm detection circuit.

To prove it's false, the alarm circuit is monitored with a data recorder. (Any signal that changes in response to the alarm is a good trigger.) With each successful trigger, the alarm circuit is separated, moving the test equipment ever closer to the faulty alarm contact. If this problem happens only once a week, the job could take a while.

14.20   Error Pulse

Computerized motion control uses a fundamental concept called *error pulse* (also known as *following error*). The error pulse (EP) simply describes the difference of where the machine "is" and where it plans to go.

When the machine is resting at its commanded *zero-point* position, the error is zero. When the servo loop is under full power, this zero-point position is fiercely defended, and if any loop error is detected, a counteracting loop correction is immediately presented. This correction results from the servo loop's continuous accounting of every position pulse. The bigger the loop error becomes, the bigger the loop correction (up to the maximum effort of which a loop is capable).

With the loop "on" and the machine at rest, the error pulse is nearly zero. (It is normally flashing between zero and a few back-and-forth pulses.) Because the distance of each *friendly* pulse is so very small, the machine is essentially being held solid at zero-point position. When the machine begins making a move, the error climbs until moving along at a constant rate where the error assumes a constant value.

A commanded move is finished, and the next move can begin when the error falls below a threshold level of error pulses. This "threshold error" is the acceptable error while still saying the machine is "in position," or close enough, anyway. Adjusting this width of *in-position* pulses higher will speed up the *block-to-block* advance in automatic running, but begins rounding off corners. It's a fundamental motion control trade-off, the advance versus the accuracy.

One way to check a servo loop is to use the error pulse data. By solving the equations of motion using some known motion parameters and fixed machine data, the error pulse readings are linked with the factory's set-up specifications.

Such calculations are capable of predicting a long list of motion loop conditions. The speed of an axis motor is calculated, evidence of a heavy mechanical load is displayed, and the overall health of a loop is suggested. Gain of the loop is also derived from additional loop constants and mathematical calculations.

Balancing the actions of two different servo loops shows up in the EP data. Error pulse readings from two mechanically identical axes, moving at the same speed, will track each other in the display. Pulse data confirms the simultaneous, balanced cutting needed to cut good circles and other such interpolations.

The newest machines have improved the response of the servo loops using feed-forward transfer functions and digital software tuning to reduce or remove traditional following error from the servo loop during running. Service investigations on these newer machines tends to exclude the direct use of error pulse data.

14.21   Alarms from Software

An excessive accumulation of error inside a servo loop generates *software* servo alarms. During a move, the error inside the loop is allowed to climb only so far. The alarm level of pulses is set by parameter values chosen at the factory. (The set-up parameter list for each axis contains these values.)

Software alarms keep coming back until the underlying cause is found and repaired. Increasing the parameter that determines the software limits only avoids and aggravates the real problem. Additional investigation and final resolution is needed.

In the case of a software servo alarm, the accumulated *lag* in the servo system exceeds the maximum soft limit. The machine was unable to get where it needed to go because of a weak loop. Everything in the loop is suspect. It may be one or a combination of a weak motor, servo, command, feedback, or simply an excessive mechanical demand. Until the weak link is found, the software alarm symptoms will continue or worsen.

The key feature of an alarm captured in software is how the alarm is quickly cleared by pressing reset. The alarm is only held in a register that is immediately cleared out by pushing the NC computer reset. There is no going around to the back, opening cabinets and looking for tripped breakers.

14.22   Test Programs

A machine spends its working lifetime cutting parts, so naturally, during an automatic cycle is when problems show up. Solving an "in program" problem takes considerable effort. Watching the program run and then alarm out helps, but the final solutions are usually found after a problem is broken down into smaller, bite-sized pieces. The segments of a main program are split into chunks of programming code called *test programs*.

The complicated main program is broken down into simple, single-operation, single-function programs to test each system on the machine. Perhaps the first test program consists of just rapid positioning moves. Next, only a programmed feed move in each axis. Another small program may just pick up an offset, or work coordinate system, and so on, until the single problem area is identified.

Writing test programs is time-consuming and demands basic programming experience. A friendly operator is the best candidate for writing and running all these piecemeal test programs. Running test programs in single block, or inserting five or ten *dummy* blocks above and below a suspected program block tests a block's lone, isolated function (dummy blocks are just blank, end-of-block characters). Many times, the solution will only surface after breaking the main program all the way down and building it back up piece by piece. Of course, since this procedure tests all the assumptions used to write the customer's main program, any silly programming mistakes will also come to light.

CNC QuizBox

14.3 Find the vibration frequency of a 1000 lb. machine on four pads.
  a.) Each pad deflects 1/32 of an inch for an applied load of 200 lb.
  b.) Deflects 1/8 of an inch for the same load.

14.6.2 The RS232 serial interface is damaged by a lightning surge. "IN" is working fine, "OUT" is dead. For common send and receive chips:
  a.) Sketch the pin-out schematics from a standard TTL data book.
  b.) Which chip is likely blown?
  c.) How can it be tested?

14.20 All three axes of an older mill are moved at 100% rapid (10m/min). During the move, the error pulse is X=4063, Y=4253 and Z=4160 pulses. If each "friendly" pulse is .001 mm/pulse,
  a.) Find the factory set-up EP value for a machine with $K_p=40s^{-1}$.
  b.) For a machine with $K_p=25 s^{-1}$.
  c.) What's the expression for motor speeds?
  d.) For a 10 mm screw and 2500 pulse/rev feedback, how fast
  are the motors turning at a 100% rapid?

15   *Overall System Features*

15.1   Introduction

The influence of industrial design gave machine tools extra curves, covers, and a sleek overall appearance; one machine even resembles a steam locomotive from the art deco period. However, a true appreciation for a machine is earned in the context of *overall* design, beyond superficial cosmetic appearance, to the all encompassing utility and function.

The in-house expertise of a machine shop makes a special impact on the demand for service, from application set-up and programming to running production and checking parts. To service these special demands of a machine shop requires attention to the area of *overall system features* for machine tools.

15.2   Thermal Expansion

A shop decides a CNC machine is cutting crazy. They cut a batch of parts and check the results on a highly precise inspection system, every batch of parts is found *out of print*.

These dismal results are trusted because the inspection room has a new, state of the art, touch probe inspection station. This coordinate measuring machine (CMM) is programmed with an automatic test routine that touches each part fifty times. The touch data automatically establishes the part geometry, and then prints out a conclusive, six-page inspection result.

Using this accurate inspection report data, new correction offsets are calculated to null the perceived CNC errors. After the new offset corrections are keyed into the control, the machine cuts another batch of test parts. When the new batch is tested, all the part tolerances have again moved; the part grew, or shrank, or something. Errors are landing all over the place.[1]

---

1 See also Section 8.10.

This "chasing offsets" begins to diminish after the machine runs extended, non-stop production.

There really is no problem with the machine or control, only the normal *thermal expansion* of the machine magnified by the accurate CMM inspection data. The long random delays between running and inspecting parts serves to exacerbate the problem.

Every machine has a range of normal thermal expansion. To document this effect a circle is cut on a machine first thing in the morning when the machine is cold. Then, after running the machine in dry production for twenty minutes, the same test circle program is cut again. On this second pass the cutter takes off new material; a few thousandth's growth is not uncommon between a cold machine and warm machine. (Alternatively, with the machine warm and switched off, an indicator set between the table and the head will begin to turn as the machine cools down.)

If the temperature of machine components change, the physical dimensions change; metal castings will expand or contract in response to heating or cooling. If the temperature rises too far, a hot casting or ball screw will eventually reach a point called *thermal saturation*[2] where the growth tapers off. The temperature of a ball screw or motor-case at this point of thermal saturation is much hotter than normal production would ever cause. (Something expensive, like covers, motors, servos, bearings, etc. will fail before the maximum saturation point is reached).

Cutting parts within print doesn't require a new machine; repeatable production will result after a reasonable *thermal equilibrium* point is established and held. Keeping the machine at this steady-state (cool) thermal equilibrium tempera-

---

2 Expansion results depend directly on machine construction. Floating screws, tensioned screws, and counteracting expansions are used. The use of linear position scales and controlled temperature barriers mitigate the error.

ture during production solves the growth problem. A twenty-minute warm-up program, first thing in the morning, brings the machine up to the production equilibrium point. Then, continuous running of production keeps it at the same spot in the thermal expansion curve. After a short lunch or pause in machining, a few dry cycles will get ready for more high-tolerance production. Simple warm up schedules will reduce marginal thermal effects so parts will cut within print.

Experienced operators add small offsets in the morning to provide good production while the machine slowly climbs up the thermal expansion curve. After the machine is in continuous production a few hours, the equilibrium point is reached and the offsets are removed. The solution to high tolerance cutting is addressed by many (at least seven) different components of error, one of which is the thermal effects.

15.3   Backlash Adjustment

Machine Side
Guy

Machine tools are tough and rugged when properly operated and maintained. They are also very unforgiving to an amateur's brute force approach. Randomly taking a wrench to the machine to "tighten that sucker up", without measuring the original backlash or adjustment results is the wrong way.

Before wrenching on a machine, the original backlash condition data is documented. Expected backlash compensation parameters are calculated and installed, and the final result tested. Larger backlash numbers (greater than twenty pulses) suggest some mechanical-based damage and repairs.

Practical tests for determining backlash include: touching off dial indicators, automatic sensor-based testing, inspecting a test cutting or performing the infamous *Gorilla* test.

The first, and most popular test for finding backlash is mounting an indicator to the machine. A short test program is written that gently feeds the machine axis down and up in three steps. Backlash data is picked up by a *tenths* indicator on the second step going down and again coming back up. In program "feeds" generally yield better, more accurate results than the manual "handle" movements.

Another popular method is simply cutting a test circle. Circles are great for physically recording the backlash during a tool pressure event. For the purpose of backlash, a circle inspects two different axes at the same time.

In one test for a machining center, the operator is asked to cut a circle with an end mill in some soft test material like aluminum. The circle center is set up as a work coordinate. After cutting a circle with a nice finish,[3] the steps left at the axes' crossover points (quadrants) are measured. Backlash data is taken directly off the test circle with an indicator gauge. Sweeping the gauge around the circle diameter fills out a circle accuracy chart (available from the OEMs). The results are then sent off to the builders for opinions and suggestions. (A qualified operator is *always* used to run a machine, gauges are

3 Machine Tool and Manufacturing Technology, Stephen F. Krar, Mario Rapisarda, Albert F. Check, Steve F. Krar, Delmar Pub, 1997.

expensive and easily damaged if accidentally driven into the table or spun in a spindle. See Gordy's story on page 41.)

During testing of a machine, the existing backlash and circular compensations (installed mainly on newer machines) can be temporarily turned off. Standard backlash parameters are zeroed out, other compensations are turned off by parameters from the OEMs. Cutting and inspecting a non-compensated circle best reflects the bare mechanicals.

Machine builders (and increasingly field service representatives) prefer *double ball bar* (DBB) sensors for checking a machine's circular movement accuracy. A DBB is a tube with a built-in position sensor. As the tube is pulled out and pushed in, the sensor sends an accurate electronic signal to measure the event. The DBB tube is securely mounted in such a way that the machine traces it around a cone shape during each full circular move.

A perfect circle will result in no elongation of the DBB tube; any flaw in the circular path causes tube elongation. The elongation data is sent to a laptop computer. The laptop is running software that converts the information into a circular accuracy chart.[4] Another more specialized test for checking circles relies on counting and displaying the feedback signals internal to a machine's own motion loops.

The last test to discuss is the infamous *Gorilla* test. In this test, an indicator is zeroed out against the table, and the table is given a snappy jerk. If the axis moves out-and-back, that's normal. If the axis moves a distance one way and stays, there's a problem. As always, the service engineer uses common sense, the machine is turned off and no crow bars or pry bars are needed, just some good footing and a snappy push on the table in each direction.

---

4   See also the footnotes in Section 1.13 and Section 10.9.

Measuring backlash is becoming an after-market business. In fact, complete accuracy checkups of machine tools are offered using touch probes, DBBs, dial gauges, laser range finding and, once again, the infamous gorilla test.

15.4   Fixtures

When final parts don't pass inspection and mispositioning is suspected, check the machine fixtures. The indelible clamping marks left on bad parts reveal the tell-tale signs of moving inside the fixture. Irregular shimmy marks or cutting grooves also indicate unsatisfactory fixture clamping.

Leaving the same part in the fixture for re-machining should not generate any fresh chips, unless something has moved. Painting the part with inspection dye between repeated looping of the test cutting helps isolate the more infrequent events of slippage or mispositioning.

The action of clamping and unclamping a finished part should not introduce a noticeable shift in the machined face or diameter. Riding an indicator against the fixture and giving a sharp tap with a rubber hammer tests the overall rigidity of the fixture. Worn out fixtures are replaced or re-surfaced. The tightening pressure applied to vices, chucks or other more automated fixturing styles[5] are carefully monitored.

15.5   Hydraulics

Adjustments to a machine's hydraulic pressure is reflected right on the gauge. By turning the pressure up, a chuck's clamping force increases. Too high and the part is crushed or distorted. Too low and the part can move, or fly out of the chuck altogether during cutting. Chuck pressure settings are often overlooked in a job's application set-up notes.

---

5 Jig and Fixture Design, 4th edition, Edward G. Hoffman, Delmar Pub, 1996.

15.6   Tooling

Complex cutting applications present detailed problems only *tooling* will resolve. The machine or control certainly seems responsible, until a new tool comes along to solve the problem. Concerns about part finish and efforts at whittling program cycle times closer to the bone are two key considerations for tooling.

Tooling recommendations are the providence of machine dealers and tooling company representatives.

15.7   Turret and Tool Changer

The ability to quickly change between cutting tools is accomplished by a *turret* on a lathe and a *tool changer* on a machining center. These electro-mechanical systems rely on carefully-aligned positioning switches and closely-timed event sequences for their operation. As they wear, mechanical changes introduce slight timing delays. As previously mentioned, these delays are unacceptable to the sequence and lead to hang-ups during an attempted tool change.

Electro-mechanical timing problems are tough to solve. While the computer systems quickly sense them, they are rarely the cause of them. An out-of-alignment switch or mechanical adjustment is the root cause. Somewhere amongst the dirty, chip-covered tool changer, the gremlin will be found. Occasionally, a bad switch or loose wiring is found straight away with a flashlight inspection.

Critical switch actions are directly monitored with voltmeters, guided by the machine-side schematics and NC-side internal diagnostics. If a switch voltage changes, but the diagnostic doesn't, then a problem exists with the interface I/O board; otherwise, and more likely, the problem is found outside the computer circuits in the machine itself.

15.8   Damage

Shops take a dim view of machine crashes. A new machine is taken out of the box and set up for production, within one month a tool holder is sheared off and the safety door is cracked. Talk about your accelerated depreciation! In retrospect, machine crashes are avoidable when a qualified operator runs a machine.

A cursory inspection betrays the signs of disasters from the past. Punctured sheet metal and broken windows cast an eye toward carelessness. A table riveted with runaway dents and chewed up fixtures raises the prospect of mechanical contribution to a machine's ongoing problems.

Certain conditions are caused by normal wear, such as backlash between the gears in a transmission, worn-out spindle bearings and the backlash between a ball screw and the positioned tables that gradually increases over the years.

When a major crash comes along, everything is screwed up in an instant. Spending exorbitant amounts for the mechanical repairs is unavoidable and often justified. Stories about machines going *crazy* are a tough sell in this business. Money saved by rushing poor set-ups and hiring low salary, untrained employees is periodically offset by the mechanical repairs that must be done to the machine and tooling when the machine crashes.

15.9   Inspection Systems

An increase in quality manufacturing has injected life into the computer inspection industry. Today, inspection results are more accurate and detailed than ever, but these improvements in inspection automation need to be carefully applied to a machine service. Machine problems described with long cryptic inspection printouts as evidence of some mysterious anomaly must be tempered with a little common sense.

Inspection is practiced in every profitable machining operation. The topic is broad and only a partial listing of topics is offered here. The reference threads provided below can help locate ample information:

CMM (Coordinate Measuring Machines)
Micrometers, Both Digital and Analog
PC Linked Inspection Systems
Diameter Checking, OD and ID
Go-NoGo Gauge Sets
Cut Open Inspections
Optical Comparators
Red Inspection Dye
Thread Pitch Inspections
Inspection Rooms
Granite Tables
Dial Gauges

There is no shortage in the literature concerning inspection systems for the machine shop.[6,7]

15.10   Sensors

To keep machine movements in order and the computer in command many different styles of motion sensors are found on machine tools. These remote positioning *switches* and *sensors* send the *completion signals* that tells the control that a motion assignment is finished on the machine.

In the course of normal machining operations, varied qualities of motion and temperature are monitored by the control: a temperature sensor detects the temperature of a motor or drive chassis, a remote magnet and sensor pair sends out motion signals to close a spindle orientation loop. Sensors refer to those electrical components capable of sensing a wide range of physical dynamics.

---

6   An article on this subject is given by Leo R. Rakowski, "CMMs boost fabricators part-inspection capabilities," MetalForming, October 1995, 56-60.
7   See also Section 10.8 and Section 1.13.

Switches send a signal when they are tripped by a metal *dog*. The gap and alignment between the switch and the dog is critical for proper signaling and machine operation. Machine tools frequently use magnetic *proximity* switches instead of the common plunger-type micro-switches.

A proximity switch is activated when a metal dog is placed in front of, or in close proximity to the switch face. The proximity switch has three wires—two for the power to make a magnetic field and a third for a high-low switching signal. When a piece of metal is placed in front of a good proximity switch, a change in the signal voltage is observed.

Switches and sensors normally get fouled up with packed cutting chips, water, and oil. A complete loss of a switch signal causes malfunction and an occasional alarm. A switch that becomes shorted will blow fuses all the way back to the main DC power source.

15.11   Fuses and Overloads

If the electric current in a circuit exceeds the maximum rating, something begins to smoke. To contain the damage from such an unexpected *over-current* event the protection of a *fuse* or *overload* is added to the circuit.

Fuses are specified for current, voltage and reaction time. Quick-blow fuses are more expensive, but give better equipment protection in the event of quick, surging currents. A fuse shows a closed circuit if it's good, or an open circuit if it's blown. (False readings are eliminated by taking a fuse outside the circuit before checking.)

The full story must be found before replacing a blown fuse. Restoring power to a shorted load causes serious additional damage. An in-rush of current can blow sparks and fire out the back of the machine leading to personal injury and expensive damage.

When a fuse blows an alarm circuit captures the event and informs the control computer. Many schemes are used to detect a blown fuse. Small *alarm fuses* have miniature internal contacts held together by a fine piece of *fuse wire*, when the wire burns, the held alarm contacts spring open. Another detection method relies on the fact that a good fuse has no voltage potential across its terminals. A blown fuse suddenly develops a voltage across it for triggering a burned fuse alarm. Finally, there are the plunger-type fuses—when they blow, a plunger pops out, tripping a nearby alarm switch. Spare fuses are available from the OEMs. All fuses and fuse alarm systems are kept at original specifications.

A different type of circuit protection is provided by the *overloads*. Current passing through an overload protected circuit is continuously monitored, and if it climbs too high, the overload's built-in alarm contacts trip a signal which shuts down the entire machine. An excessive current inside the overload causes small wire loops to glow red hot, heating up an adjacent bimetallic strip. When the strip gets too hot, it curls back and opens up the tiny overload alarm contacts. Heavy current is not opened by the overload, only the gold plated alarm contacts are opened.

After the overload cools down, it is manually reset. Overloads can be adjusted for different levels of current, so during an overload replacement, the original current settings are recorded and transferred to the new unit being installed. When the reason the overload current became excessive is analyzed, a bad motor is often found; rarely is the overload itself the cause.

15.12   Ground Fault Protection

The main circuit breaker mounted on the back of the machine is familiar to anyone who works around machine tools. These three-phase breakers are specified for load current and a lesser known rating called the *ground fault detection* current. These two ratings are important protection systems for both the operators and the machine itself.

Normally, a balanced three-phase current flows through the main breaker and into the back of the machine. If current levels climb above the breaker's rated load current, it trips off.

If any stray current flows into the chassis of the machine, the breaker trips again, but this time half-way down to the ground fault position. One cause of these deadly and dangerous ground fault currents is water-based coolant seeping into the core of an AC motor, which allows a small current path into the chassis. Like the ground fault test button found on the plug next to the bathroom sink, the main breaker on a machine tool has a small red or yellow fault-circuit test button.

CNC QuizBox

15.2 Find the expansion of a 3 m cylindrical bar of steel. The bar diameter is 40 mm.
  a.) Over a temperature range of 0 C° to 100 C°.
  b.) Over a temperature range of 30 C° to 31 Celsius degrees.
  c.) Over a temperature range of 68 °F to 70 degrees Fahrenheit.

15.3.1 Backlash compensation is often misunderstood. Every system has backlash as a normal condition. Answer the following:
  a.) How is a CNC system compensated for its backlash?
  b.) What triggers backlash compensation?
  c.) What are the units of backlash?
  d.) How is the backlash compensation turned off?

15.3.2 The amount of existing mechanical backlash is found before any adjustments, mechanical or control-side. Answer the following:
  a.) Why is backlash checked with feed, rather than with handle moves?
  b.) Make a procedure for checking the linear backlash on a turning center. Include: i.) A sketch showing axis movement and indicator placement.
       ii.) A short incremental test program (with spindle off).
       iii.) A blank chart for recording three trials of test data.
  c.) Repeat Part b.) This time for checking the rotary backlash on a machining center's rotary 4th axis.

15.10 Starting with 12 Vdc, how much does the voltage drop:
  a.) Across a closed set of switch contacts?
  b.) Across 10 feet of 16 gauge copper wire.

3rd Tool

Problem Solving

Introduction

Professional service engineers have the goal of fixing highly detailed and complicated machinery. And if not fix, they improve the situation without causing any new problems. The role of shop management is to understand how a proposed solution proves a problem is solved before accepting and signing off on the engineer's work. In both cases, reliable problem solving plays a role in the mission.

Successful repairs show a balanced use of service skills. By themselves the knowledge and diagnostic tools from Parts One and Two can occasionally fix a problem, but the crafty use of all "three skills" together solve the toughest problems. Less successful attempts show a mysterious lack, or imbalance in these three skill areas.

Solving a problem starts with breaking it down, or *separating* it into the three different skill types. As long as the necessary skills are identified and supplied, a problem remains solvable. This is the standard used to explain problem solving in this book. *Separate* the problem and apply the necessary tools.

Certain problem separations are performed at the end-users. Additional separations are carried ahead by dealers and independents, finally the highly-specialized separations are left for the builders or other experts who have the specialized knowledge and diagnostic techniques.

The more frequent problems are solved almost immediately—a single separation suggested by the problem's description does the trick. (Oh yeah, heard that, now try this—it works.) Complex problems take successive double, triple, or multiple separations. Multiple separations are akin to a growing flow chart: if one type of separation becomes too complicated, an alternate set of less demanding separations branch out around the roadblock.

This is reminiscent of the *scientific method*—build a chain of trusted separations using accurate experimentation until an overwhelming force of evidence is built up from which the problem can't escape.

The "easy" single separation problems begin to propagate after solving the more detailed multi-separation problems. The tougher the problem, the better. This is done in the same way that algebra skills can become elementary after solving some intense calculus problems. The three principles discussed in this book are practiced and learned *together* by solving the tough problems.

A structure for separating problems is gradually built-up over the next three chapters. Chapter Sixteen deliberately solves the single separation problems using a single skill from the often discussed ToolBox of "three skills." In Chapter Seventeen a "game plan" technique begins work on the multi-separation problems. The final chapter discusses five types of representative problems often encountered in the field.

When a level of confidence and consistency is maintained from start to finish, the inevitable roadblocks and setbacks just serve to challenge the service. New attacks from other angles are developed until finally, a ridiculously obvious solution arrives. It is uncanny how simple these answers really are—after they're found.

16 Three Skills for Service

16.1   Introduction

A down machine leads the technician (or engineer) to seek the fastest service approach. Picking up the phone to contact someone who has a quick answer is the preferred method. Phone answers solve the problem without lengthy, expensive service investigations and naturally impress the customer—when they work.

Other quick answers arise from those service skills used repetitively and subsequently memorized—not a fresh challenge to an engineer's knowledge, diagnostic, or problem solving ability. Quick answers are repetitive and well-worn. Inexperienced service engineers, without first-hand knowledge of a particular machine, can *rattle off* a pretty good answer if the problem is a fairly common one.

Solving every service problem in this "rattle off" style is an ever-hopeful dream around the service network. The latest strides in expert computer systems are improving the percentages, but still the need for good, experienced servicemen isn't likely to wane anytime soon.

This chapter illustrates quick, one-dimensional service skills by solving a few simplified examples using single *knowledge, diagnostic* and *problem solving* answers. These *three skills* are applied separately to create "rattle-off" quick answers. (In all the upcoming examples, notice the letters to the left of each simple answer. They signify which skill set is applied.)

The tougher service problems won't yield to this "rattle-off" approach, they are both stubborn and oblivious to simple rote memory. The tougher cases are solved with a skill combination, or multi-dimensional approach outlined in the final two chapters.

16.2   Problem Solving **(PS)**

During a field service repair the knowledge and diagnostic information is distilled into a final solution using the discipline of problem solving. The purpose of this section is to illustrate the singular skill of problem solving. No deep insight into CNCs, just the common sense approach to solving problems. Like the old doctor's joke, when a patient reports, 'the problem starts every time I do this,' the easy answer is, 'well, don't do that.' *Problem solving* skills are unconnected to the other more formal service skills. They are more fun and creative, less structured and serious.

Consider a warm-up question: "Did anything in the shop change around the time a problem started?" If the answer is "yes," and a light bulb goes on, send the bill—problem solved.

Q. My lathe's communication interface is dead!

(PS)   **Has this function ever worked?**

Yes. In fact, before we installed our new PC computer, both the lathe and mill talked to our old system for years without any problem.

(PS)   **Your new system won't talk to either machine?**

Well, the new mill is receiving OK, but the old lathe is getting nothing—that's the problem!

(PS)   **Tried the mill's good system on the bad lathe?**

No. . . Wait a minute, I'll bet it's—the cable! See we have a new fork-lift driver here. . . I'll call back, thanks.

The "glue" for successful service is problem solving. The subject of creative problem solving is challenging in principle and equally challenging in practice.[1] Observation, thought and experience leads to good problem solving.

16.3 Knowledge Tools (K)

In practice, machine tool service is often just a lengthy process of elimination. Recall Part One where the discussion focused on the network of service, personal safety issues and the machine itself. It was pointed out that additional details are available, and often necessary.

In the case of a long, drawn out repair, a service engineer's collection of previous knowledge eventually runs dry. During a long job, it will happen. Naturally an experienced engineer gets further than the rookies, but either way during a long job, it will happen. When it does, pushing the job forward demands finding that next piece of information. The key idea will either be learned (and learned the hard way), or picked up from somewhere else (which is the easy way).

Q. My lathe's communication interface is dead!

(K)   **You mean your serial RS232 interface?**

---

1 Creative problem solving is a big business for the consultants. An interesting link for a small to medium sized machine shop is www.halcyon.com at extension www.halcyon.com/qcss/, an interesting "QC newsletter" is updated monthly.

Uh. . . yes.

(**PS**)   **Has it ever worked?**

Yes, right up until our fork-lift ran over the computer cable this morning.

(**PS**)   **How can I help?**

How are the send and receive wires connected to the NC-side, 25-pin connector? And why is my desktop computer using a different 9-pin connector?

(**K**)   **Pin 2 is send and pin 3 is receive. The NC-side send signal is hooked to the PC-side receive signal. They cross somewhere in the cable. Would you like a schematic faxed over for both the 25- and 9-pin style connectors?**

Yes, as soon as possible. Thanks.

School is one place that learning is taught. Beyond reading and arithmetic is the underlying experience of learning to learn. Solving the long grueling homework problems instills the student with new ideas and concepts. Calling up a smart lab partner, however, gets the quick answers. Quality service depends heavily on finding service engineers who know the good people to call.

Rattling off an incredible solution to some obscure machine problem is easy when the symptoms fit the template of a known problem. The two-week jobs that boil down to a single loose wire or misplaced jumper are never forgotten and are quickly rattled off to the next amazed contestant.

During a normal service job the rattled-off good guesses often turn out to be wrong, so before time is wasted chasing after the easy solutions that never quite fit, time is spent up front collecting sound diagnostic data.

16.4   Diagnostic Tools (D)

When an NC alarm says, "cable is unhooked" and sure enough, it's found hanging—then bingo! No other efforts are needed. By running simple diagnostics on the machine the data needed to quickly rule out (or rule in) a previously known solution is found.

> Q. My lathe's communication interface is dead!

(PS)   **Has it ever worked?**
Well, after I installed a new communication
cable today, the program data started sporadically
losing characters during transfer.

(D)   **Just for testing purposes, try slowing down
the communication baud rate. Does that help?**
I'll try it out. . . hang on. . . .
Yes, when the baud is below 2400, it's fine.

(K)   **Sounds like the new cable changed the signal-to-
noise (S/N) ratio. Is the new cable different?**
The new cable is routed upstairs, to the new computer
in the front office.

(K)   **Oh, longer cable, more signal attenuation and
frequency smearing. Can you shorten it up?**
I'll try. Thanks.

In this example, diagnostic checking determined the maximum rate of error-free communication. The system functions at 2400 baud. In practice, higher communication speeds are more useful and reasonably demanded. The maximum length of cable is spelled out in the communication standards, if it's too long, it won't run to specification.

These three preceding examples were dispatched using elements of the three skills. In the next chapter the three skills are again applied, but this time as building blocks in a simple game plan structure.

CNC QuizBox

16.1 What are the three basic tools or skills in the standard CNC ToolBox?

16.4 "Rattle off" a good service response using each set of skills.
a.) Q. What is the machine's stroke in Z-axis?
    i.) (K)
    ii.) (D)
    iii.) (PS)

b.) Q. Why can't I change a parameter?
    i.) (K)
    ii.) (D)
    iii.) (PS)

c.) Q. What is the control type for my machine?
    i.) (K)
    ii.) (D)
    iii.) (PS)

17 The Game Plan

17.1   Introduction

When a sense of wandering has crept into a "tougher-than-expected" repair it's time for the service engineer to come up with a sensible *game plan*. A game plan reduces overly complex problems into an approachable set of realistic tasks. These tasks are completed using the skills described in the *action logo*. Piece by piece a service plan is carried out.

Steps for building and executing a service game plan are reviewed in the next four sections. In each section "Gordy" the serviceman works on a customers problem.

17.2   Understand the Problem (Observe Problem)

Simple observation is the first step a service engineer uses to formulate a game plan. Everyone who has seen the problem is asked to fully explain or demonstrate the trouble. After reaching a clear understanding of the problem a *checklist* of necessary tools and areas for testing is compiled.

"First, Observe the Problem"

> A problem has been reported! Every time a machine is turned on, the main breaker trips back off. I need to go on site to check out this machine.

The customer reports the main breaker trips off at power up. The machine has a mysterious problem— I better push in the emergency stop push button before turning on the main power.

Normally, the main breaker is turned on by flipping up on the switch. On this machine, the breaker is in a "halfway" down position. That's interesting—it needs to be pushed down first, then it clicks back up to the normal "on" position.

The main breaker is energized and the machine is now receiving main power. Now, turn on the control computer. So far, everything seems normal, the display screens all came up. Well, there's no apparent problem with first power, just the resident Servo Off and E-Stop status alarms.

Nothing unusual happened until releasing the E-Stop and going for second power. Just that quick—the machine blinks out. Around back, the main breaker trips back down to that strange "halfway down" position.

17.3   Make a Plan (Separate the Problem)

The overly complex service tasks are reduced down to smaller bite sized pieces. A problem broken down into enough parts will lead to the solution. This process is called *separating the problem*. Reasonable separations of a problem are suggested by the symptoms of the problem.

As an example, assume the "right answer" is found after separating the problem six times. This implies six tests were needed to determine that six others were reasonable. Out of

275

twelve tests, only one demonstrated the final "answer." It takes a lot of spirit to be "wrong" eleven times before finding the answer, but in hindsight, each of those eleven steps was just a necessary part of the repair.

Passing along known separations to a more qualified person often saves them time, and gives everyone involved a chance to build knowledge from the successful separations they contributed. A faulty separation slows the process, is ultimately uncovered as such, and ignored.

The more advanced the service skills the more efficient and potentially beautiful the separations. In any case, a feasible set of the skills outlined in the knowledge and diagnostics sections are applied to the case.

The character of a finished repair is reflected in the efficiency that separations arrive at the answer. If all possible separations are imagined, many paths will eventually lead to a solution. Of course, one path is elegant, but the others are fine if they somehow do the trick. The short direct road, the long winding road, and the road leading through *hell* are all on the same map.

"Make a Plan"

Time for making a game plan. I'll return to the notes of the problem, put the observations in order and label all the necessary skills.

OBSERVATION 1.)

Machine is fine at first power and the normal set of alarms are observed.

OBSERVATION 2.)

The breaker trips off the instant second power is added.

OBSERVATION 3.)

The problem causes the breaker to trip half-way down.

(K 1)   Reading the manufacturer's data sheet on the circuit breaker explains that two types of protection are offered. The first type is excessive current protection, and the second is ground fault protection. The half-trip means a ground fault condition was detected in the main power source.

(D 1)   Try checking the main power at the circuit breaker with a meter.

OBSERVATION 4.)

Apparently, the machine is having "ground fault" at the instant of second power-up.

"Separate the Problem"

(K2)   The circuit breaker's data sheet further explains that the ground fault alarm protects the people in physical contact with the machine chassis. Stray leakage currents anywhere out on the machine triggers the protection. The ground fault detection scheme used inside the breaker is complicated and sensitive. It is somehow related to either an imbalance in the three phase main power or a "ground leakage."

(D2)   Maybe the circuit breaker is bad! This ground fault detection current they talk about can't really be checked. Unless a better idea comes along, take the machine off line and change out the breaker, or find someone else who can check this mysterious "ground fault current."

(PS 1) Wait a minute, the ground fault condition doesn't exist at first power. Mysterious "ground fault" currents only flow when the heavy circuits are activated. Assume, for the time being, that the breaker is good. That means the "leakage" is in the second power circuit. But where?

(K3)   When I look at the electrical drawings for the second power circuit, it shows a big contactor distributing the big power out to each servo driver. From there, the servo drivers pass this heavy current on to each motor. A bad servo, motor, or wiring is a possible "ground fault" suspect. That is a lot of things to check out!

(D3)   If a circuit powers up, without tripping the ground fault, it is not causing the problem. Conversely, when the bad circuit is powered up, the breaker will trip. Rather than swapping parts around, I will start by inspecting the entire servo system (in a power off condition) with a good flashlight.

(D4)   If necessary, components in the servo loop's main circuit can be replaced one-by-one, but this is an expensive and unattractive option.

(PS2)   The faulty circuit must be energized to cause the ground fault.

17.4   Use the Plan (Collect the Data)

With the problem observed, broken down, and a few good service options identified, it's time for Gordy to decide on a preliminary service plan. At a glance, the actions needed to advance the repair are documented. A carefully written-out list of observations document the problem and points out when something is missing or may need more detail. The available "tools" are selected, the plan is carried out and the results are recorded.

The right people for each stage in the repair are identified. (The inherent complexity of a modern CNC often requires the service of experts, each keen in their own specialty.) So far in Gordy's problem the background knowledge has been gathered from reading the manuals and calling around to the service network.

Gordy will execute the plan and call on the necessary experts. If a plan doesn't reach a final solution, the initial results are used to keep modifying and building up the process.

"Use the Plan"

I will review the initial game plan and execute only the easiest and safest steps. At each step I will collect more data on the problem. I will continue in this fashion until the problem is solved, or the collected data suggests it's time for better-qualified outside assistance.

My plan is summarized and grouped into the tools suggested by the initial observations.

The Knowledge Tools:

K1. Data sheet reviews fault (not entirely explained).

K2. Ground fault is a key safety interlock (complicated).

K3. Schematics for servo power circuit are extensive.

The Diagnostic Tools:

D1. Check incoming power for phase balance.

D2. Swap breaker or hire a special diagnostic service.

D3. Inspect servo system (powered off, using only a flashlight).

D4. Swap parts (dangerous and expensive, should check more).

The Problem Solving Tools:

PS1. Ground fault occurs at second power only.

PS2. Faulty circuit must be energized to cause "ground fault."

"Collect Test Data"

I will collect more data on the customer's problem. I pick two items from the list, checking the main power (D1) and inspecting the second power circuit (D3).

279

COLLECTING DATA (D1)

D1. **Check incoming power for phase balance.**

Checking the machine's main power is dangerous so I call over the plants qualified electrician to help. I make a small drawing of the proposed test which contains the check points, and list all the test equipment requirements. A digital voltmeter with high tension probes is selected to do my testing. The selected meter is first powered up to gain familiarity with its operation. Both test probes are in good shape. Everything seems in order.

Now the power outside the machine is shut off at the taps and locked out by the plant electrician. The test probes are briefly touched to the test circuit to make sure it's off. The AC voltage reads zero on the voltmeter, indicating that the main shut off outside the machine is doing its job. A padlock on the main tap cut-off switch prevents it from accidentally getting switched back on.

The voltmeter probes are securely installed to the circuit under test. The meter is placed in plain view of the operator's power-on button and then taped down in place with strong black electrical tape. Everything in the test set-up is secured.

The padlock at the wall is taken off by the plant electrician and the machine is briefly re-energized. The voltmeter registers the proper phase-to-phase line voltage and the data is written down in the test chart. Power is then completely shut down again, the test set-up is moved by the plant electrician to the next test point until all the test readings are separately displayed and recorded.

The main voltage to the machine is found to be within the OEM's specifications, so there is no apparent problem here.

COLLECTING DATA (D3)

D3. **Inspect servo system (power off, with a flashlight).**

Again, the main power to the machine is locked out at the wall. It's time to visibly check the second power circuit using the machine's schematic and a good flashlight. Each branch, of each servo loop, is located and inspected, looking for bad connections, burn marks, or anything unusual.

Inside the cabinets, everything looks clean and tight. The main servo contactor isn't welded shut and doesn't show any sign of excessive arcing. Each servo-unit looks the same, and the wiring is found to be good and tight.

Out on the machine, the wires to each motor are proving hard to find. They are mostly hidden by a slew of machine covers and sealed cable conduits, and everything is very oily. Using a couple of shop towels, the motor casings and connected cables for each servo axis are cleaned.

Remembering that the power is still safely locked out, the motors and associated wiring are all carefully inspected with the flashlight. This process continues at length for each of the axis motors on the machine.

Low and behold, out on the second axis motor, a quantity of coolant has spilled out of the loosened feedback cable. Using a mirror and a flashlight, it appears that water is coming out of the motor.

17.5   Look Back (Using Hindsight)

When a possible solution is presented, look back and verify the results. Then, consider the other ramifications of the solution.

"Using Hindsight"

How did water get inside the motor? One problem is apparent, but how did it happen?

Clearly, the motor and cable are both wet and need to be dried out and possibly replaced. Replacing the motor on this type of machine is too big of a job for me. It's time to make some phone calls—a call to the control builder lines up a spare motor and servo-unit for overnight delivery. A call to the machine builder secures a serviceman for tomorrow who is qualified to change the motor on this type of machine.

The overnight parts and the engineer both arrive the next morning, on schedule. The question is, "how did coolant water get inside the motor."

The first thing the machine builder's engineer does is to hook up the wiring from the wet motor, to the new motor, which is sitting in a packing crate balanced on the machine. The engineer is very careful to block up the weight of the machine that is no longer supported by the disconnected motor (without big stable blocks, the machine could come crashing down at power-up).

The machine powers up, and for the first time in three days, the main breaker didn't trip off at second power. Satisfied, the serviceman shuts everything down and locks out the power, and begins the lengthy, unpleasant task of removing the wet motor.

Apparently, this is not a new problem. In fact, the machine-side engineer explains that a small drainage hole buried deep inside the machine gets plugged up. Water backs up and finally submerges the motor shaft, letting water seep around the shaft seal and into the main motor cavity. From there, it ran the entire length of the motor and out the sealed electrical cables in the rear. A ground fault alarm was triggered by stray current flowing through the wet motor chassis.

The coolant drainage hole on the machine is drilled out oversize. The bill for one day's labor, plus the price of the motor, is written up. A recommendation for checking the coolant drainage hole every few months is given in the report. After getting a signature of approval, the machine-side engineer quickly departs with a hearty handshake. The machine is up and running again thanks to the coordinated effort of the maintenance network.

In the next chapter, the scheme of solving problems within the game plan structure is tailored to the five categories of problems often experienced in the field.

CNC QuizBox

17.2 Assuming a person's body resistance is 400 Ω and a person can just survive a current time of 20 mA–s. How quickly does a 200 V circuit have to be shut down for a connected person to survive?

17.4 A large machining center runs a routine for rigid tapping. The threads suddenly start getting torn up during the tapping cycle. Propose a plan for investigating the problem. Include all three skill sets in the plan. Prepare several testing charts and methods.

18 *Solving Problems*

18.1 Introduction

When confronted with a new problem, there is one key question, "How often does it happen?" The answer affects the overall problem solving process.

If a problem happens all the time and is in plain view the straightforward approach is taken. However, if a problem only occurs once every six months during the night shift, an obvious adjustment to the normal problem solving approach is needed.

When more and more symptoms are observed, a problem moves closer to solution. On the other hand, a problem having no observable symptoms and only a small dose of second hand conjecture becomes a tough nut to crack. Both extremes are expected in the daily course of machine tool service.

Five progressive stages are assigned in this final chapter to cover the range between the always there and the never seen "Loch Ness-Monster" type problems. At each arbitrary *rate of symptom* stage a service engineer will adjust the service techniques depending on the problem at hand.

18.2 Always Problems

The machine was working fine one minute, and in the next minute, it went down. Now, the problem happens every time and is easily observed. This kind of problem normally doesn't require a lengthy service visit—just a few phone-in suggestions from the OEM before everything quickly returns to the status quo.

18.3 Intermittent Problems

Initially, an urgent call goes out to get everything lined up in case the machine should quit altogether. With the service safety net in place, a busy shop then decides to keep on running until the current job is finished. The shop knows how to keep the machine running, until finally, the machine is shut down and released for the unwelcome service investigation.

With intermittent problems, everyone who is consulted has a couple of good ideas, but nobody knows with complete certainty what will fix the problem. Many reasonable possibilities exist. Only a full on-site service investigation can sort it out with real certainty.

18.4   Very Intermittent Problems

"If I can't see it, I can't fix it!" is a common refrain after a service engineer waits around for two days and no evidence of a problem is seen. The service is ended and five minutes after driving away the problem shows up again plain as day. The attitude begins to lean towards replacing the machine "if this damn mystery" is allowed to linger on much longer.

There are generally two ways to go—one way is to set up clever diagnostic traps to catch the problem, and the second is to relentlessly sniff around for better clues and deduce the

Better diagnostics and better clues are the keys to solving the very intermittent type problems.

problem directly. These are the kind of problems that provide the rare and memorable "real-world" experience for a service engineer.

During a long stretch while waiting for a problem to surface, there is plenty of time for meticulously chasing down each and every clue in the shop. Elaborate diagnostic setups are devised and exhaustively tested. Time is wisely spent studying the problem, staying busy and filling time. In this way the big-ticket labor charges will be justified with results.

Everyone who has seen the problem is interviewed. If an operator witnessed the problem the service engineer will hang around for the night shift and have a nice chat. The machine operators spend eight hours a day running the machine and will notice when anything changes. They are a prime source for missing details concerning the problem.

A problem may only occur once or twice in a day (or week). This is a clear prescription for using high-speed data recorders. During the lull between events, it is made absolutely certain a proposed diagnostic test will work. What a shame, having a service engineer standing around for two days drinking coffee, and then, when the long-awaited alarm trigger finally arrives—the event is missed because of a dangling probe, misconnection, or some other service error. The opportunity is lost, maybe forever.

After getting a shop's best story the service engineer will decide on the best game plan. The ultimate goal with "very intermittent" problems is devising a shrewd method to recreate the problem at will. Only this will triumphantly assure everyone that the problem is found and forever solved.

18.5   Not Really Problems (Out of Specifications)

Most experienced service people know a simple test program that will alarm out (and possibly damage) a machine. There is nothing fundamentally wrong with the machine; there is only a brutal test program that pushes the machine outside the envelope of design specifications.

This test program is one example of the "not really" or *out of specification* type problems. These surely look like problems and act like problems, but the machine is actually fine—it's just being asked to do something that is beyond its capabilities. The role of service in these cases is to interpret the benchmark limits of the machine's original factory specifications. If the expectation for a machine's operation are outside the benchmark, either the expectations must change or

another machine must be acquired that has the expected specifications.

Setting up the benchmark specifications for a machine
tool relies on first capturing advanced diagnostic data and then
applying equally advanced data interpretation knowledge.

This type of service will acknowledge a careful, balanced, open-minded service approach.

Maybe there is a deep, serious problem which the factory
design team will ultimately need to know about. A sincere
attitude and belief by the service engineer that there could *really* be a problem must begin the service. By collecting more
data, the service engineer slowly builds a picture that tells the
true, unbiased, and in all honesty, unknown story.

The truthful history of the system is recalled—how over
many years the problem being suggested has *never* been demonstrated. There is no place for unsubstantiated explanations—everyone must stand by the well documented
paradigms. They serve to motivate finding the "real" reason
behind a problem.

Good data solves the hazy political arguments. Fax a
copy of indisputable service results to all the protagonists.

18.6   "Do Not Touch" Problems

The final type of problems that come up in the field are
the no-win situations. If things start out with meaningless descriptions like, "The machine is crazy!" or "The last guy spent
a week and couldn't figure it out," maybe the problem is outside the scope of service.

A service job requires that something is really broken and
can be fixed. The observation stage of the repair is used to

determine if a service problem exists. A detailed history of the problem from all those involved may determine something has been brewing outside the scope of service. Demands to start swapping parts without reason are answered with statements like, "It can't do that, every position pulse is counted". . . "This machine performs up to specification". . . "There is an alarm for that condition," and a personal favorite "This application may need some re-evaluating."

The three skills leading to successful service of CNC machines are the same skills mysteriously missing when unsuccessful events are reported. Top service draws upon a balance of skills including the contributions of knowledge from Part One, the diagnostics from Part Two and the art of solving the problem in Part Three.

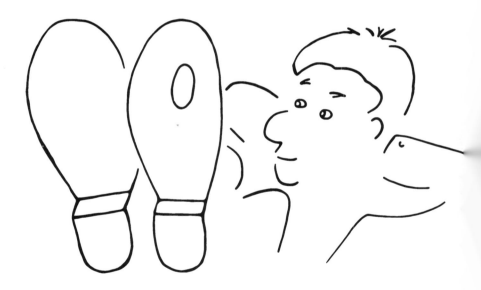

Job Well Done, Time to Relax

Conclusion

The OEMs develop quality CNC service techniques through extensive classroom training and by sending their engineers to hundreds of shops in the field, increasing the level of service difficulty as they progress. After two or three years of wide-ranging field experience an engineer reaches a critical mass where a majority of problems are confidently understood and handled. The OEMs have the expertise to get the job done right, because after all, they are the original designers.

The picture below is a simple road map for how OEMs service their own machines. Three general service skills: knowledge, diagnostics and problem solving, work together to accomplish the task of service.

To some degree, all people involved with CNCs possess these skills. Solving the service problems, especially the tougher problems, with success and consistency demands a balanced application of these three disciplines. The size of service toolbox carried by any service organization is characterized by the level of performance when applying these skills.

The first discipline, **knowledge**, is treated merely as a tool. Some knowledge tools are found in books, others result from experience, while the rest are available if the right questions are asked of the right people. In machine tool service, generating new knowledge only reinvents the wheel—someone else has already found and paid for the same answer.

The second discipline, **diagnostics**, is the skill of collecting pertinent data to understand a machine's trouble. Diagnostics are often fast moving and dynamic, reflecting a fleeting machine status or change. The first type of diagnostics are specifically built into the machine: the alarm codes and computer based monitors. Another type of diagnostics are those collected from outside sources like inspection data, keen observations and the results of portable hand-carried test equipment.

Finally, both the knowledge and diagnostic information is distilled into a final solution using the third discipline, **problem solving**. In this section, a problem is solved using rote memory. Easy enough! But what happens when confronted by the unknown? Then, problem solving becomes a most interesting and challenging tool.

The subject of machine tool service is vast. Running footnotes are used throughout this book to refer the reader to a wide list of outside references. The appendices in the back show how to get help and information from the major OEMs. Included are a standard list of factory publications and a longer list of specific service documents. These items are surprisingly easy to find, especially if you own one of the builders machines.

Though the answers are not given in the text, do not shy away from the exercises at the end of each chapter. They cover important technical areas that would not quite fit in the book. Searching through the many references will yield the answers.

Readers world-wide can quickly locate references for this book or find out more about the QuizBox exercises at the internet web site for Aero Publishing: *www.cncbookshelf.com* Send along any new links or other books you come across for machine tool service to this site. They are posted to help everyone in their search for better CNC service.

Appendices

Appendix 1: Factory Publications

The machine makers supply "owners manuals" to explain programming, operating, machine- and control-side service. The standards for these books vary between builders; some provide updated readable versions, while others hardly bothered to publish more than a single operating manual.

A listing of the standard publications supplied for a lathe or a machining center are given with brief topic summaries. Not every machine receives every manual. After checking the choices, locate the items needed, or found missing. Beware, the purchase price for factory published books is very high.

Machine Specific Publications

Diagnostic List:

Specific assignments for finding status of "lamps and switches" on machine. List of level-one signals. Explains meaning of machine sequence parameters. Helps clear up sequence alarms and interlocks. Often combined with machine operator's manual.

Mechanical Drawings:

Core assembly and machine adjustments. Sectional views of motor couplings, bearings and mountings. Machine parts list.

Machine Operator's Manual:

List of machine capabilities and specifications. Instructions for operating tool magazine and other machine specific systems, like part pallets, tool holders and touch tool setting.

Machine Sales Brochures:

Describes machine in colorful brochure. Lists all basic and optional features for the machine. May include sales comparisons between builders. Brief explanations for the more popular machine options. Examples include high-speed cutting and fast spindles.

PC Ladder:

A standard written form of the machine sequence software that explains the I/O-Signal mapping used to link a specific control computer to a specific machine configuration. Powerful service tool in the right hands.

Control Specific Publications

Computer Communication Manual:

Brochure explaining the standard and optional features for serial communication with control computer. Includes all related parameter settings for desired protocol selections.

Connecting Manual:

Manual supplied to machine builder by control builder explaining all connection schemes for using control. Helpful on service.

Control Operator's Manual:

Contains programming instructions and diagnostics for computer. Lists all the control functions available. See machine operator's manual for those specifically selected for a machine application.

Elementary Schematic Drawing:

Connector names and pin numbers of wiring between servo, spindle and control PC-Boards. Shows power-on and power distribution circuits. Wiring from I/O-Boards to machine switches shown here, or in a separately prepared set of machine-side schematics.

Internal Schematic Drawing:

A three part set of drawings. Lists of tiny board level components soldered to the PC-Boards, with their electrical description and manufacture. A skeleton drawing that links this parts list to all the board level locations. And finally, the actual schematic drawings for all the board level wiring.

Maintenance Manual:

Listing of alarm codes and trouble shooting checks. Listing of internal, level-two diagnostics. Description of PC-Boards.

Sequence Design Manual:

Manual supplied to machine builder by control builder for guiding the writing and loading of the machine-side sequence software.

Servo Application Manual:

Shows elementary schematics and mounting instructions for the axis servo amplifiers. Contains servo alarms and control signals with their full explanation. All servo setting procedures and running specifications. Axis ratings for motor power, encoder details and application matching of motor and drive.

Servo Schematics:

Internal wiring of the drive transistors, firing circuits and I/O circuits.

Shows elementary schematics and mounting instructions for the spindle drive units. Contains spindle alarms and control signals, with their full explanation. All spindle setting procedures and running specifications. Spindle ratings for motor power, encoder details and application matching of motor and drive.

Spindle Drive Schematics:

Internal wiring of the main semi-conductor modules, firing signals and control I/O-Signals.

Upgrading and Optional Functions:

New revisions for the original control operator's manual. Includes newly developed improvements for the control. The operation and application of new G-codes, alarms and software displays.

Appendix 2: Specialized Documents

1st Tool–*Knowledge*

Maintenance Network

1.11 Procedure for faxing a problem if phone is too busy.
1.12 Procedure for requesting service.
1.14 Schedule and content of classes. (See Appendix Three)

Computer Numerical Control

3.4 Explanation for new canned cycles not in manual.
3.5 Exchange procedure for PC-Boards inside CPU rack.
3.6 Regeneration procedure to clear computer memory.
     Factory original set-up parameter list.
3.7 Description of latest NC software updates and items changed.
3.8 List of sequence software versions and ladder diagram.

Electric Powers

4.1 Full specification for the machine's main power service.
4.2 Current rating for installing a step down transformer.
4.5 Schematics for the DC power supply.
4.8 Circuits involved in power-on sequence.

Servo Loop

5.4 Summary of DC servo-unit settings by model type.
     Exchange procedure of DC servo by model type.
     Elementary schematic for servo-unit.
5.5 Summary of AC servo-unit settings by model type.
     Exchange procedure of AC servo by model type.
     Elementary schematic for servo-unit.
5.6 Feedback unit specifications based on model type.
5.7 Encoder Specifications based on model type.

Spindle Loop

6.4 Elementary schematic of DC Spindle Drive showing main and
     control circuits.
6.5 Exchange procedure for DC drive and boards.
6.6 Speed generator specifications, based on model type.
     Field and armature resistance and characteristic curves.
     Preventative maintenance for DC motor brushes.

2nd Tool–*Diagnostics*

Servo Loop

The External Servo Diagnostics

The Internal Servo Diagnostics

Spindle Loop

The External Spindle Diagnostics

The Internal Spindle Diagnostics

Computer Numerical Control

The External CNC Diagnostics

The Internal CNC Diagnostics

Appendix 3: On-Site Training Index

Target training opportunities from the OEMs, dealers, independents and professional trade societies using this listing.

Index

A

Absolute Positioning, 100, 136, 140
    Feedback Unit, 110, 100
    See also Position
AC Motor. See Induction Motor
Action Logo, 13, 274
After Market
    Consultants, 168, 30
    Options, 166
    Parts, 168, 24
    See also Internet
Air Cutting, 56
Alarm
    False, 243
    Hardware, 65, 13 , 140, 242
    Reset, 243, 260, 95
    Servo, 94, 244
    Software, 64 , 137, 138, 140
    Spindle, 120, 125, 224
    Sequence, See Ladder
    Toolchanger, 154
    Trigger, 212
Application Software. See CAD/CAM
ASCII, 145
Axis, 81
    Freedom, 81, 139
    Interpolation, 64, 81, 153, 245
    Labeling, 80, 97
    Linear, 80 , 151
    Pure, 81
    Rotary, 81, 164
    See also Servo, and Spindle

B

Backlash
    Adjusting, 141, 152
    Finding, 122, 253
    See also Compensation
Ball Screw, 139, 142, 150
    Pitch, 151

Base Block, 94, 120
Base Speed, 122
Baud Rate, 145, 163
    See also RS232
Bit, 145
Board Level, 228
    Automatic Testing, 228
    Serial Data, 52
    Signature Analysis, 214
Brushes
    Servo Motor, 104 , 210
    Worn, 108, 212
    See also Spindle Motor
Buffer. See Remote Buffer
Bus Voltage, 123, 124, 217
    See also Dangers
Byte, 145

C

CAD/CAM, 167
    Contour Programs, 82,145,163,167
    G Code Filtering, 58
Capacitors, 40, 42
    See also Dangers
Central Processing Unit (CPU), 48
Characteristics
    Signal to Noise Ratio, 271
    Spindle Motor, 213
    See also Specification
Check Pin, 195, 228
Circle Cutting, 140
    Accuracy Chart, 153, 253, 254
    Compensations, 140, 253
    Double Bar Ball, 254, 180
    Non-compensated, 153, 254
    Quadrants, 253
Circuit Protection
    Fuse, 72 , 259
    Ground Fault, 260
    In Rush, 259
    Overload, 259
Cleaning
    Cooling Fans, 132, 143
    Motors, 107, 128
    Safety, 107, 231
    Swarf, 48, 157
Closed Loop, 62, 82, 89, 98

311

Order Form

Additional copies of this book are available by telephone, fax, mail order and internet through our service and fulfillment company, PSI Fulfillment. They provide immediate order processing and shipping for all titles from Aero Publishing.

Telephone orders: Call toll free: **1(800) 965-6328.**
Have your Visa or Mastercard ready.

Fax a purchase order: (512) 288-5055

Internet: www.cncbookshelf.com

Postal Orders: PSI Fulfillment, c/o Aero Publishing, 8803 Tara Lane, Austin, TX 78737. USA
Fill out a copy of this form and send to address above. Allow 4-6 weeks for delivery.

Company Name:_____

Name:_____

Address:_____

City:_____State:_____Zip:_____

ISBN	Author/Title	Price	Qty.	Total
0-965431-47-9	Nelson/The CNC ToolBox	$54.95	___	_____

Shipping: $4.50 for the first book and $1.50 for each additional book.   _____

Sales Tax: Please add 5.5% for orders shipped to (WI) addresses.   _____

Payment: _____

☐ Check (Made out to PSI Fulfillment)
☐ Credit Card: ☐ Visa ☐ Mastercard

Card number:_____

Name on card:_____Exp. date:___/___

Standard discount schedules are available. Prices subject to change.
Foreign orders add $20.00 handling charge.

Call toll-free to order now.